Genetically Modified Plants and Beyond

Edited by Idah Sithole Niang

Published in London, United Kingdom

IntechOpen

Supporting open minds since 2005

Genetically Modified Plants and Beyond
http://dx.doi.org/10.5772/intechopen.95212
Edited by Idah Sithole Niang

Contributors
Sinan Meriç, Alp Ayan, Tamer Gümüş, Çimen Atak, Robert Redden, Jennifer Thomson, Mohammedsaani Shehasen, Melaku Addisu, Girma Hailemichael, Selvi Subramanian, Muthulakshmi Chellamuthu, Kokiladevi Eswaran, Prince Emmanuel Norman, Daniel K. Dzidzienyo, Kumba Yannah Karim, Aloysius A. Beah, Ashutosh Kumar, Banshidhar, Priyanka Jaiswal, Harmeet Singh Janeja, Richard Mundembe, Endale Gebre Kedisso, Karim Maredia, Nicholas Barro, Lilian Chimphepo, Tahani Elagib, Rose Suniso Maxwell Gidado, Benard Oloo, Ruth Mbabazi

First published in London, United Kingdom, 2022 by IntechOpen
IntechOpen is the global imprint of INTECHOPEN LIMITED, registered in England and Wales, registration number: 11086078, 5 Princes Gate Court, London, SW7 2QJ, United Kingdom
Printed in Croatia

British Library Cataloguing-in-Publication Data
A catalogue record for this book is available from the British Library

Additional hard and PDF copies can be obtained from orders@intechopen.com

Genetically Modified Plants and Beyond
Edited by Idah Sithole Niang
p. cm.
Print ISBN 978-1-83969-875-0
Online ISBN 978-1-83969-876-7
eBook (PDF) ISBN 978-1-83969-877-4

For EU product safety concerns:
IN TECH d.o.o., Prolaz Marije Krucifikse Kozulić 3, 51000 Rijeka, Croatia, info@intechopen.com or visit our website at intechopen.com.

Meet the editor

Idah Sithole Niang graduated from Michigan State University, USA. She is a professor in the Department of Biotechnology and Biochemistry, University of Zimbabwe, Africa, where she teaches Molecular Biology and Biotechnology and conducts research on exploiting the biotechnological potential of bacteriophages and bacterial and fungal endophytes. She maintains a keen interest in biotechnology and biosafety issues. She has edited and co-edited several books and is the former chairperson of the University of Zimbabwe Publications Committee.

Contents

Preface XI

Section 1
Transformation Methodology and Biotechnology Applications 1

Chapter 1 3
Genetic Engineering for Oil Modification
by Muthulakshmi Chellamuthu, Kokiladevi Eswaran and Selvi Subramanian

Chapter 2 25
Genetic Modification and Application in Cassava, Sweetpotato and Yams
by Prince Emmanuel Norman, Daniel K. Dzidzienyo, Kumba Yannah Karim and Aloysius A. Beah

Chapter 3 49
Next Generation of Transgenic Plants: From Farming to Pharming
by Alp Ayan, Sinan Meriç, Tamer Gümüş and Çimen Atak

Chapter 4 83
Site-Specific Recombination and the Development of Cisgenic Plants
by Richard Mundembe

Chapter 5 95
Genetic Conservation and Importance of Ginger in Ethiopia
by Girma Hailemichael, Mohammedsani Zakir and Melaku Addisu

Chapter 6 107
Crop Biotechnology and Smallholder Farmers in Africa
by Endale Gebre Kedisso, Nicolas Barro, Lilian Chimphepo, Tahani Elagib, Rose Gidado, Ruth Mbabazi, Bernard Oloo and Karim Maredia

Section 2
Biosafety Framework and Consumer Perceptions 131

Chapter 7 133
Proposed Revision of the National Gene Technology Scheme for Australia
by Robert Redden

Chapter 8 145
Biosafety Aspects of Genetically Modified Crops
by Ashutosh Kumar, Banshidhar, Priyanka Jaiswal and Harmeet Singh Janeja

Chapter 9 151
GM Crops: The West versus the Rest
by Jennifer Thomson

Preface

As the global debate and controversy continue to rage around genetically modified crops (GM crops), and various reasons are proffered for public concerns about these crops, the science is stepping up to the plate to help address some of these concerns. The book contains nine chapters that speak to the development and utility of these crops. In particular, it discusses the root and tuber crops, cassava (*Manihot esculenta* Crantz), sweet potato (*Ipomoer batatas*), and yams (*Dioscorea* spp.), as well as ginger and oilseed crops including *Jatropha*. The book also discusses in detail the use of plants to develop pharmaceuticals. Moreover, it examines the biosafety regulation of these crops, including regulatory frameworks form Australia and India, and differences in perceptions between the West and the rest of the world to strategies for removing selectable marker genes.

Chapter 1 outlines the biology of candidate oilseed crops and the genetic modification and genome editing approaches used for overall improvement of oils and their profiles. Furthermore, these approaches are geared towards addressing the ever-growing demand for oil-based products in human diets as well as health, pharmaceutical, and other industrial applications. The improved yields and healthier profiles also help address this growing demand without necessarily increasing the area/land under cultivation. This is a welcome development given that globally the amount of arable land is decreasing, while the demand to feed an ever-growing population is increasing. This demand also speaks to newer ways of farming with less water used per unit area. The chapter also refers to the utility of *Jatropha* as a resource for biofuels. Its diverse attributes make it an attractive crop for genetic improvement and industrial application as well. The range and choice of oilseed crops discussed in this chapter are representative of the desired product range and various applications.

Chapter 2 speaks on the latest developments in molecular breeding and gene manipulation for the three species of cassava (*Manihot esculenta* Crantz), sweet potato (*Ipomoer batatas*), and yams (*Dioscorea* spp.). The authors also delve into the future prospects of increasing efficiencies and broadening the repertoire of new and existing selectable marker genes. The chapter also covers the applications and potential benefits of genetic modification in breeding selected outcrossing root and tuber crops, thus circumventing the challenges faced by breeders using conventional breeding approaches.

Chapter 3 deals with the development of recombinant proteins in plants. This is a rapidly growing area recognized for its product safety, cost-effectiveness, scalability, diversity, and numerous ways to rapidly produce recombinant proteins. Plant systems have significant advantages and out-compete both animal and yeast recombinant protein production systems. They are amenable to large-scale production of biomolecules such as antibodies and therapeutic proteins. Molecular pharming with transgenic plant systems has added advantages over other production systems that are fraught with contamination challenges. These systems are equally adept at post-translational modifications such as glycosylation, and the product can be stored in user-friendly forms such as seeds and leaves, among other forms of tissue.

Chapter 4 is on generating cisgenic plants using original or sequences from related species to generate marker-free plants, which is fast becoming a reality. The chapter also provides a rich background to the four mechanisms that govern site-specific recombination, the *in vitro* assays that are used for each, and the advantages and disadvantages of each approach. This chapter also serves as a quick reference guide for teachers, regulators, skeptics, and students alike. Whether these plants escape the rigorous regulatory assessment used on GM crops or whether they are also going to be subjected to stiff regulation remains to be seen. As more and more examples and case-by-case examples are generated, perhaps that dawn is nearer than currently envisaged. All sides continue to watch this space with bated breath.

Chapter 5 is an exhaustive treatise on the topic of ginger, discussing breeding and genetics, crop selection, propagation, marketing, challenges, opportunities, and future prospects. Being fully aware of the criteria for breeding and crop selection as an initial step to rendering ginger amenable to genetic manipulation is a great achievement. This is a comprehensive reference guide for teachers, researchers, students of ginger, farmers, and traders alike, which makes it a tremendous resource.

Chapter 6 presents case studies of biotechnology applications and progress made in six countries in Sub-Saharan Africa including Burkina Faso, Ethiopia, Kenya, Malawi, Nigeria, Sudan, and Uganda to address biotic, abiotic constraints, and malnutrition that smallholder farmers face every day on the continent. The case examples are long-term studies of more than twenty years.

Chapter 7 is on the proposed revision of the national gene technology scheme (NGTS) for Australia, which includes the regulation of gene editing (GE) technologies. The chapter comes with a recommendation that an education campaign should be launched to ensure wider distribution and understanding by the general public, including school children. Indeed, this call goes beyond the borders of Australia as other countries could learn from the Australian example and benefit from the lessons learned and key issues that must be considered when dealing with such matters.

Chapter 8 deals with the GM regulatory framework for India, including the development, use, import, and export (transboundary movement) of such crops. The chapter outlines how the parent ministry, the Ministry of Environment & Forests, is involved in capacity building and supervision of the various committees under its purview, highlighting fruitful areas of engagement for other countries to glean from.

The final Chapter 9 deals with unpacking the trajectory of the controversy between perceptions held by people in the West and the rest of the world and the consequent knock-on effect it has on continents like Africa and the rest of the world. Some of these perceptions arise from myths so rife that politicians, decision makers, and indeed citizens from these countries embrace them as fact. The author provides an in-depth analysis of why these myths are taken as fact and concludes by highlighting the fact that these mistakes must be taken as lessons learnt in order to engage more fruitfully as countries begin to address newer technologies such as genome editing.

Idah Sithole Niang
Department of Biotechnology and Biochemistry,
University of Zimbabwe,
Harare, Zimbabwe

Section 1

Transformation Methodology and Biotechnology Applications

Chapter 1

Genetic Engineering for Oil Modification

Muthulakshmi Chellamuthu, Kokiladevi Eswaran and Selvi Subramanian

Abstract

Genetic manipulation is a strong tool for modifying crops to produce a considerably wider range of valuable products which gratifies human health benefits and industrial needs. Oilseed crops can be modified both for improving the existing lipid products and engineering novel lipid products. Global demand for vegetable oils is rising as a result of rising per capita consumption of oil in our dietary habits and its use in biofuels. There are numerous potential markets for renewable, carbon-neutral, 'eco-friendly' oil-based compounds produced by crops as substitutes for non-renewable petroleum products. Existing oil crops, on the other hand, have limited fatty acid compositions, making them unsuitable for use as industrial feedstocks. As a result, increasing oil output is necessary to fulfill rising demand. Increasing the oil content of oilseed crops is one way to increase oil yield without expanding the area under cultivation. Besides, the pharmaceutical and nutraceutical values of oilseed crops are being improved by genetic engineering techniques. This chapter addresses the current state of the art gene manipulation strategies followed in oilseed crops for oil modification to fulfill the growing human needs.

Keywords: oil quality, yield, essential fatty acids

1. Introduction

During India's green revolution in the mid-twentieth century, the use of agrochemicals and high-yielding crop types established through traditional plant breeding procedures resulted in a major increase in crop productivity [1]. Conventional plant breeding can no longer meet the ever-increasing global food demand. Food insecurity and malnutrition are two of the most major threats to human health today, claiming the lives of millions of people in poor countries. To stay healthy, we need to eat a range of meals that contain all of the needed nutrients, as well as those that provide health advantages beyond basic nutrition [2]. It is now way to encourage sustainable farming approaches for increasing crop output while preserving all natural resource to the greatest extent possible [3]. Agricultural biotechnology is proven to be a valuable addition to traditional ways for addressing the global need for high-quality food. We now have access to large gene pools that may be utilized to confer desirable characteristics in economically significant crops thanks to modern plant biotechnology technologies. Crop varieties that are genetically modified (GM) can help us satisfy the demand for high-yielding, nutritionally balanced, biotic and abiotic stress tolerant crops [4]. Oilseed crop adoption has increased significantly in recent decades as a result of high demand for human consumption and industry interest. The composition of the seed oils, which are composed of a broad group of fatty acids with six predominant types and other unusual fatty acids produced by wild plant species with chain lengths ranging from 8 to 24 carbons, such as 16 or 18 carbon palmitic, stearic, oleic, linoleic, and linolenic acids, and 12 carbon lauric acid. In this study, a review of the major advances in genetic improvement of oilseed crops is provided, beginning with omics to understand metabolic routes and identify key genes in seed oil production and progression to use modern biotechnology. Genetic engineering is a new breeding technique (NBT) that has enabled the functional study of genes with potential applications. The important advancements in plant genetic improvement using current biotechnology, with an emphasis on oilseed crops such as *Sesamum indicum*, *Arachis hypogaea*, *Carthamus tinctorius* and *Jatropha curcas* are discussed in the following sections.

2. Genetic modified crops vs conventional breeding

In several countries, GM crops created by adding genes for greater agronomic performance and/or enhanced nutrition are commercially grown. The source of the DNA utilized to develop the GM crop has a significant impact on the rigor of the food safety assessment. If the DNA comes from an edible plant, the regulatory process prior to commercialization will be streamlined, and customer acceptance will improve [5]. Crops that have been traditionally bred and those that have been genetically modified through various methods of gene transfer technologies are the results of genetic changes. Both conventional breeding and GM technologies have the potential to alter an organism's genetic makeup in terms of DNA sequences and gene order. However, compared to traditional breeding, where thousands of uncharacterized genes of an organism may be involved, the quantity of genetic modifications brought about by GM technology is limited and clearly documented. Furthermore, GM crops are the result of very specific and targeted gene modifications, with well-defined end products like as proteins, metabolites, and phenotypes [6].

3. *Sesamum indicum*

The genus of *Sesamum* belongs to the clade eudicots; order Lamiales and Pedaliaceae family and broadly grown species around the world [7]. The genus *Sesamum* contains 36 species including 22 species from African continent, seven is found commonly in Asia and Africa, five in Asia and one species in Brazil and Greek island. Most of the wild species of *Sesamum* originated in the African continent however the crop has been domesticated from its wild relative species *S. malabaricum* native to south Asia [8]. Sesame harbors a vast range of diversity and adaptation to various environments and it was recorded with long-term natural and artificial selections [9]. The percentage content of other fatty acids like oleic acid, linoleic acid, palmitic acid, erudic acid are 36%, 30%, 9%, 0.8% respectively. These are the major fatty acids present in sesame. Linolenic acid (omega 3 fatty acid) content is in very trace amounts in sesame seeds. The percentage content of poly unsaturated fatty acids ranges from 30.9 to 52.5%, it shows very large variation in their germplasm (**Figure 1**) [10].

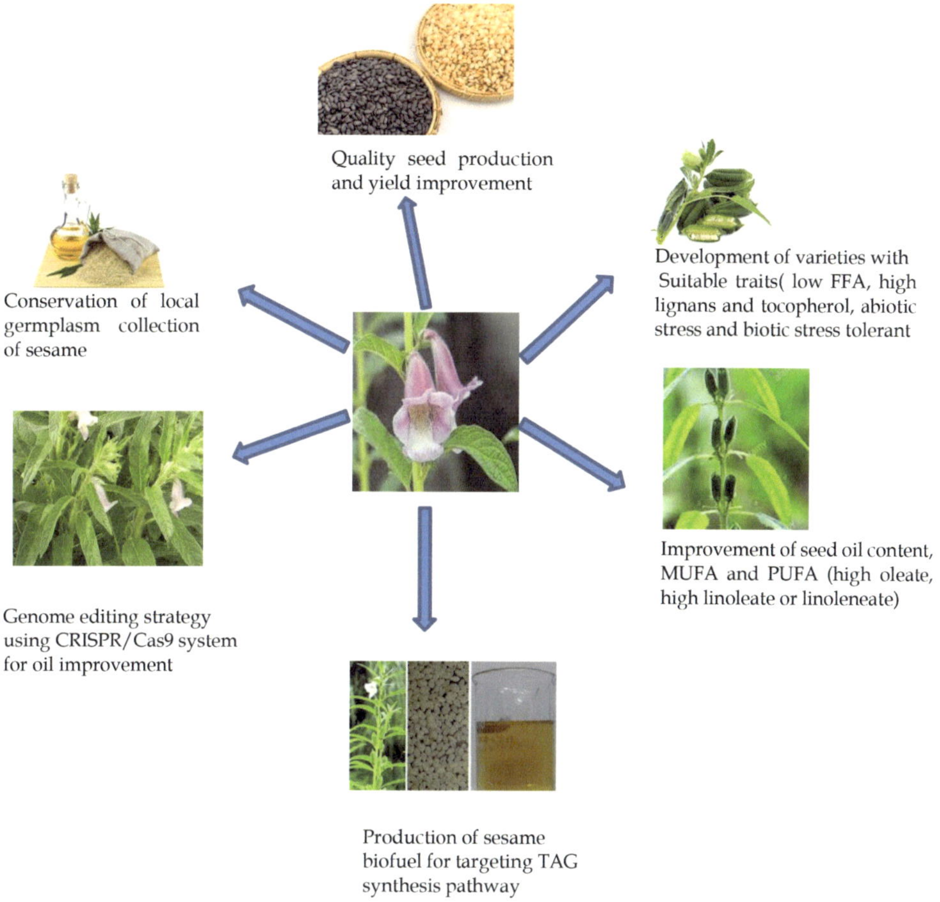

Figure 1.
Future directions and strategies for enhancing sesame oil yield and improvement.

4. Nutritional value of sesame

Sesame not only contains protein, carbohydrates, poly unsaturated fatty acids, it also contains the lignans, phytosterols, phytates and tocopherols. They keep on maintaining the oil quality level in long shelf life time by preventing the oxidative rancidity [11]. The combination of these compounds is mainly responsible for the good oxidative stability of the sesame seed oil [12]. The antioxidant property of the oil aids in preventing the degenerative diseases like cancer, cardiovascular disease, atherosclerosis and the process of aging [13].

The major desmethylsterols present in sesame seed oil are β-sitosterol, campasterol, stigma sterol, Δ-5 and avenasterol [14]. Sesame oil also contain some enzymes such as Protex 7 L, Alcalase 2.4 L, Viscozyme L, Natuzyme and kemzyme. Among those enzymes Alcalase is found in large amounts in sesame [15]. These enzymes are mainly used for aqueous oil extraction process which is an alternative for solvent extraction. An Enzyme-assisted aqueous extraction (EAAE) process which is used to recover the high-quality protein for human consumption [16].

5. Sesame breeding

Plant breeding allows the successful management of existing genetic diversity as well as the development of new ones in order to achieve desired traits. There is different type of breeding approaches which is employed for genetic improvement of sesame varying from plant selection, hybrid development and molecular breeding. In conventional breeding the choice of parental lines and development of sesame types with desired characters is attained through pedigree selection from segregating generations [17]. Plant selection is vital for increasing seed yield and development of novel sesame varieties [18]. Several phenotypic traits are useful for determining selection criteria such as number of capsules, branching, biomass, harvest index which reveals positive correlation with sesame seed yield [19]. Hybridization is one of the frequently used techniques in conventional breeding technique. Combination of desired traits with different plant lines can be achieved through cross-pollination. Cytoplasmic male sterility lines in sesame were developed by hybridizing *S. indicum* with its wild relative *S. malabaricum*. Many hybrids exhibited high heterosis for oil content, seed yield and number of capsules per plant [20]. Mutation breeding involves induction of new genetic variability through spontaneous or artificial mutagens either chemical or physical. Sesame mutants have been developed for desirable traits for quality, seed color, higher yield, plant architecture and larger seed size [21]. The gamma ray induced mutants were developed with improved plant growth having determinate growth habit, resistance to *Fusarium* blight, improved oil quality with higher oleic acid and low linoleic acid content [22].

6. Genetic improvement of sesame

Sesame breeding uses a variety of novel ways, including genetic engineering, to overcome the disadvantages of traditional breeding. Sesame's resistance to current biotechnology makes it difficult to use. Furthermore, various researchers have tried a variety of ways and media to create callus tissue [23]. Cotyledons, root, hypocotyl segments and sub apical hypocotyl of seedlings were all successful in somatic embryogenesis [24]. In addition, the efficient micro propagation mechanism for sesame conservation and multiplication has been upgraded. This is useful for genetic transformation, reproductive growth, and other tissue culture research. The genetic transformation of sesame by Agrobacterium has been reported, however the transformation frequency is low. High-frequency sesame transformation techniques recently yielded high regeneration and transformation frequency of 57.33% and 42.66%, respectively, for sesame [25–28]. Current crop breeding approaches will not be sufficient to meet the ever-increasing population's demands for food security and nutrition. To speed agricultural genetic improvement, 5G breeding tactics such as genome assembly, germplasm characterization, gene function identification, genomic breeding methodologies, and gene editing technologies have been proposed [29]. Genomic tools and methodologies for phenotype discovery and molecular breeding are provided by genome assembly. A gene expression, proteome, metabolome, and epigenome maps are essential. Researchers from the Chinese Academy of Agricultural Science's Oil Crops Research Institute and other institutions have successfully created a high-quality sesame genome. Two landraces (S. indicum cv. Baizhima and Mishuozhima) and three modern cultivars (S. indicum var. Zhongzhi 13, Yuzhi 11, and Swetha) have genome assembly presently available, providing a significant tool for comparative genomic analysis and gene identification [30]. In seeds of *Nicotiana tabacum*, expression of sesame plastidial FAD7 desaturase modified with endoplasmic reticulum targeting and retention signals increases a-linolenic acid accumulation. The expression of the modified sesame ω-3 desaturase raises the a-linolenic acid concentration in the seeds of transgenic tobacco plants by 4.78–6.77%, while lowering the linoleic acid level. The findings suggested that the engineered plastidial ω-3 desaturase from sesame has the potential to influence the profile of a-linolenic acid in tobacco plants by shifting the carbon flux away from linoleic acid, and thus it could be used in a genetic engineering strategy to increase a-linolenic acid levels [31]. Increases in oil content and seed weight were seen when sesame DGAT1 was overexpressed in many lines of *Arabidopsis* thaliana 'Col 0' [32]. Through a genetic engineering technique, the *Fusarium moniliforme* 12/15 bifunctional desaturase gene was used to increase the omega 3 fatty acid content of sesame (Unpublished data). Yeast is a great model for studying lipid production. The oil accumulation and functional characterization of the sesame DGAT and PDAT genes were studied using a yeast H1246 oil synthesis defective mutant [33]. In order to improve the oil quality, another study examined the co-expression of DGAT1 and PDAT1 genes with omega 3 desaturase genes in a yeast expression system (Unpublished data). Sesame transformation research using Agrobacterium to assess the biodiesel potential of transgenic sesame plants showed an increased TAG content by 10% when PDAT1 and FAD3 were combined in a transgenic construct [34].

7. CRISPR/Cas system for oil production and quality improvement in sesame

Although some of the candidates for oil characteristics are extremely suggestive, they are still suspected causal genes. The creation of several biparental populations from well-designed crosses will increase mapping resolution, allow epistatic interactions to be identified, and allow the development of new germplasm with improved phenotypic performance. To validate the impacts of these candidate genes and their functional variations for the connections underpinning oil characteristics, functional genomics approaches such as genetic transformation and genome-editing technologies using the CRISPR/Cas system are needed. Sesame genes for oil production and quality are likely to play major roles in other closely related oilseed species (for example, sunflower), allowing researchers to search for genes with similar functions. This work in sesame may provide unique knowledge and guiding examples for continuing genetic investigations for oilseed crops with more complicated genomes [35].

8. *Arachis hypogaea*

Groundnut (*A. hypogaea*), often known as peanut, is a major oil, food, and feed legume crop farmed in more than a hundred nations. Groundnut is prized for its high calories content, which comes from oil (48–50%) and protein (25–28%) in the kernels. From 100 g of kernels, they supply 564 kcal of energy. Furthermore, groundnut kernels are high in mono-unsaturated fatty acids and contain several health-promoting substances such as minerals, antioxidants, and vitamins. They include antioxidants such as p-coumaric acid and resveratrol, as well as Vitamin E and a variety of B-complex vitamins and minerals such as thiamin, pantothenic acid, vitamin B-6, folates, and niacin. Groundnut is a good source of bioactive polyphenols, flavonoids, and isoflavones in the diet. Groundnut and groundnut-based products can be promoted as nutritional foods to combat energy, protein, and micronutrient deficiency among the poor due to their high nutritional value [36].

9. Genetic manipulation in peanut

Genetic transformation can make it easier to introduce possible candidate genes into plants for controlling a variety of crop-improvement features. Transformation technology paved the way for key genes to be transferred into the peanut genome for improved resistance to fungal, viral, and other pests, drought, and salinity, as well as the silencing of undesired genes and improved nutrient uptake. Transgenic peanuts with the human *Bcl-xL* gene expressed in their genome demonstrated high tolerance to oxidative and salt stresses [37]. By compartmentalizing Na^+ ions in the vacuoles, overexpression of the *AtNHX1* gene in peanut (a vacuolar Na^+/H^+ anti-portar) increased resistance to extreme salinity and water deprivation [38]. Under field settings, *PDH45*, a pea DNA helicase similar to eiF4A, displayed abiotic stress tolerance and increased peanut productivity at T3 generation [39]. In another study, transgenic peanuts expressing the *AtNAC2* and *MuNAC4* (NAM, ATAF, and CUC) transcription factors conferred drought, moisture stress, and salinity tolerance, as well as increased crop output [40, 41]. The list of genetically modified traits were shown in **Table 1**.

Trait Evaluation	Reference
Abiotic stress tolerance	
Empirical approach for drought tolerance	[39]
Transpiration efficiency	[40]
SPAD chlorophyll meter reading (SCMR), specific leaf area (SLA), carbon isotope discrimination [and thus water-use efficiency (WUE)]	[41, 42]
High temperature	[43, 44]
Salinity	[45, 46]
Aluminum toxicity	[47, 48]
Biotic and abiotic stress	[34]
Photosynthetic rate, stomata conductance and higher transpiration rate under limited water conditions	[49]
Resistant to higher salt and water	[35]
Drought stress tolerance	[50–52]
Salt and oxidative stress	[53]
Improved water/drought stress tolerance	[37]
Drought and salinity tolerance	[38]
Water deficit stress	[54]
Nitrogen fixation tolerant to soil drying	[55]
Fungal resistant varieties	
Resistance against *C. personata*	[56]
Resistance against the late leaf spot disease	[57]
Resistance against three fungal pathogens	[58]
Late leaf spot disease	[59]
Resistance towards *C. arachidicola and A. flavus*	[60]
Virus resistant varieties	
Resistance to TSV infection	[61]

Trait Evaluation	Reference
Resistance to PBNV	[62]
Resistance to PStV	[63]
Resistance to TSWV	[64]
Insect/pest resistance varieties	
Resistance against *S. litura*	[65]
Insecticidal activity against *H. parallela*	[66]
S.minor resistance	[67, 68]
S. litura resistance	[69]
Resistance against lepidopteron insect larvae of lesser crosstalk borer	[70]
Tomato spotted wilt virus (TSWV)	[71]
Vaccine producing varieties	
Edible vaccine against *Helicobacter pylori*	[72]
Blue tongue outer coat protein that comprises the neutralizing the epitopes	[73]
Allergen silencing varieties	
Produced hypo allergenic peanut by silencing Arah 2 and Arah 6 genes	[34]
Alleviated peanut allergy	[74]
Genome editing technology	
CRISPR/Cas9 mutagenesis of *FAD2* genes	[75]
TALEN mutagenesis of fatty acid desaturase 2	[76]

Table 1.
List of genetically modified peanut traits.

10. Genome editing in peanut

The CRISPR/Cas9 system is based on the prokaryotic type II CRISPR system, which was derived from a gene editing mechanism in bacteria. It's a relatively new technique that allows researchers to change the DNA of an organism for the sake of research. Breeders can use this technique to add, remove, or modify genetic materials at a precise point in the genome. In comparison to ZFNs and TALENs, the CRISPR/Cas9 system stands out for its ease of use, efficiency, and low cost, as well as its capacity to target multiple genes [77]. Gene-editing technology has a lot of potential for improving peanut oleic acid. The first gene editing in the model plants *Arabidopsis thaliana* and *Nicotiana benthamina* using CRISPR/Cas9 was reported in 2013 [71, 78]. Since then, it has been widely used in many plant species for gene function research, and its current widespread use in crop breeding shows promise for future breeding programmes. The limited specificity of sgRNA in CRISPR/Cas9 may result in off-target DNA sequences. An unanticipated or undesirable mutation will occur in the organism's genome as a result of this consequence. Despite the fact that cas9 nickase was developed to decrease the off-target effect, improvement is still required [79]. The use of gene editing techniques makes the creation of double-strand breaks in chromosomes much faster than using conventional breeding techniques. Double stranded breaks (DSBs) can be utilized to deliver targeted disease resistance and genome alterations to improve agronomic parameters such as yield and nutritional content by harnessing the natural cellular DNA repair process [80, 81]. To characterize the functions of peanut AhNFR1 and AhNFR5 genes in the nodulation symbiosis, researchers used hairy root-mediated CRISPR knockout. The findings not only confirmed that using CRISPR/Cas9 in combination with a hairy root transformation system is a quick way to characterize gene functions in roots, but they also improved our understanding of the role of the NFR genes in peanut nodulation [82].

11. *Carthamus tinctorius*

Safflower (*C. tinctorius*) is a versatile crop that can be grown in the tropics and subtropics in semi-arid climates [83]. Safflower seed cakes provide a high protein source, feed for animals and birds, and traditional medicine. Safflower oil is rich in oleic as well as linoleic acid [84]. In addition to these traditional applications, safflower is increasingly being used to synthesize transgenic goods, including pharmaceutical ginseng, human insulin, and apolipoprotein [85]. Safflower has evolved into a platform for industrial food production due to its low outcrossing rate and weediness, distinctive appearance from other oilseed crops, and excellent agronomic characteristics, such as the taproot architecture that allows it to access subsoil water reserves [86]. It has been commercially successful to genetically modify safflower, but there is no detailed description of how to generate and analyze transgenic T1 plants in the public domain. The lack of reliable regeneration of transgenic T1 progeny in safflower has enormous implications for this economically-important plant's capacity to be used as a high yielding industrial crop. Safflower is undoubtedly a challenging crop to genetically engineer, and there is substantial literature that describes limitations of tissue culture techniques for safflower [87, 88].

12. Crop improvement of safflower

Although safflowers produce some of the healthiest oils for human consumption, their agronomical features of drought resistance and arid region adaptation prevent them from becoming a major crop. The lack of oil and yield is due to its low oil content and susceptibility to diseases and insect pests as compared to other oilseed crops like canola and cotton. Plant breeding has produced a range of cultivars that have different fatty acid profile oils, quantities, and quality, with the primary use being edible and industrial oils, along with a minor use as bird seed. This comprises specialized oils with high -linoleic acid (gamma-linoleic acid, GLA) and higher tocopherol content, which are thought to offer health benefits. Safflower oil offers potential in the biofuel industry as well as foundation for pharmaceutical manufacture in GM safflower seed [85, 89–91]. Current Australian varieties contain up to 42% oil whereas in United States have developed cultivars with oil content levels ranging from 45 to 55% [92]. In India, the most prevalent breeding approach for safflower cultivar production is choice from indigenous varieties, and multiple germplasm lines with required qualities have been created. Through selection and/or hybridization with local lines, this material can then be used for breeding in other countries. Safflower cultivars were produced in the twentieth century in the United States, Canada, and Argentina, using material imported from India, Russia, and Turkey [93]. The most complicated variables in safflower are seed yield and oil content, and selection for these traits is impeded by substantial genetic-environmental interactions. For the production of hybrid safflower plants, dominant and recessive genetic male sterility (GMS), cytoplasmic male sterility (CMS), and temperature sensitive genetic male sterility (TGMS) systems have been established. In India, GMS safflower lines (including spiny and non-spiny flowered lines) with a 20–25% increase in seed and oil output are available. In India, CMS and TGMS lines are also commercially accessible. Despite the development of hybrid safflower production technologies and the testing of hybrids, practical production of hybrid safflower is still a long way off [94, 95].

13. Fatty acid modification in safflower

Oil seed crops like safflower are primarily grown for their high-quality edible oil. All safflower seeds contain fatty acids including linoleic acid, stearic acid, and palmitic acid. Safflower lines had improved fatty acid compositions comprising reduced palmitic acid, reduced stearic acid, and high to very high linoleic and oleic acids with reduced saturated fatty acids such as palmitic and stearic acids [96]. In research published by [97], the al allele has been linked to a defective fatty acid desaturase (FAD2-1) (fatty acid desaturase) enzyme in microsomes. These vegetable oils contain a higher level of oleic acid and are more nutritionally beneficial [98, 99]. Saffola 517, a high-oleic-oil type, and Saffola 555, a linoleic-oil variant, were both introduced to Australia from the United States. Traditional breeding and genetic modification were used to create HO cultivars. Biodiesel, lubricants, and hydraulic oils are all items that require strong oxidative stability; therefore HO vegetable oils with high oxidative stability have non-food applications or prospective industrial usage [100]. GLA is a crucial fatty acid needed by the body, derived from linoleic acid by means of delta-6-desaturase in the endoplasmic reticulum. The oil content, viability, or fitness of high GLA lines is invariable and heritable across generations. SonovaTM 400, a nutritional supplement containing GLA extracted from GM safflower, has received FDA approval for use. It has been shown in clinical trials that GLA can be useful to treat eczema and various types of cancer [85]. The high level of oleic acid (75–85%) found in some safflower cultivars is ideal for food use but not for industrial use due to the extremely high level of purity required. Potential industrial applications for high oxidative stability HO vegetable oils include biodiesel, lubricants, hydraulic oils, and oleo chemical applications. The oxidative stability of oil extracted from super high oleic (SHO) safflower was significantly improved when compared to the high oleic acid cultivar S317, which contained over 93% and 75.4% oleic acid, respectively. The seed-specific RNAi-silencing of the FATB and FAD2.2 genes, which are responsible for the release of saturated medium-chain fatty acids and the desaturation of oleic acid to linoleic acid, respectively, was used to create the SHO safflower [101]. Bio fortification of safflower, an oil seed crop was genetically modified to improve the ALA content. In safflower, accumulation of Linoleic acid is higher which an immediate precursor of ALA. Hence, FAD3 isolated from *A. thaliana* driven under seed specific promoter isolated from *Glycine max* is transformed through agrobacterium mediated transformation to increase the ALA content. The vector used for cloning is pCAMBIA2300. The transformed seeds contained about 1.34–18.2 mg of ALA per gram dry weight of the seeds. Thus, it proves that fatty acid desaturase can increase the accumulation of ALA content in plants [102].

14. *Jatropha curcas*

Jatropha is a second-generation biofuel resource that is prized for its high oil content, low seed cost, ease of land reclamation, and adaptability to a variety of marginal and semi-marginal areas [103]. The extensive potential of this plant, as well as the many uses of different plant components, has made cultivation of this species highly profitable [104]. Because fossil fuels constitute a significant danger to energy security and have negative environmental consequences, efforts are underway to partially replace fossil fuels with biofuels. The high oil content of up to 50% of its seeds, which can be easily processed to partially or completely replace petroleum-based diesel fuel, has recently attracted interest [105, 106]. Jatropha is a non-food crop, which distinguishes it from the fuel vs. food debate. It has a flash point of 235°C and a calorific value of 39.63 MJ kg^{-1}, making it appropriate for use as a biofuel. Jatropha oil has a similar composition as peanut, palm, and corn oil, with 45.79% oleic acid (18:1), 32.27% linoleic acid (18:2), 13.37% palmitic acid (16:0), and 5.43% stearic acid (18:0). Jatropha is second only to oil palm in terms of oil production per hectare, which encourages its planting around the world. To mitigate financial risk, jatropha farmers have reportedly avoided cultivating the crop on marginal and ruinous lands, but this is no longer possible [107–109]. Because jatropha is widely available in India, it can be used as an alternative energy source to ensure the country's energy security. By 2020, India plans to increase biodiesel production and replace 20% of diesel usage. Depending on the potential yield of the plant types and additional improvement projects, the area required to accomplish this substitution aim ranges from 4.24 to 66.98 million hectares (Mha). Because of the vast amount of open wastelands in India, this goal is achievable. The CSIR-CSMCRI is well-known around the world for its work on Jatropha elite accessions selection, cultivation, genetic enhancement, and biodiesel production (**Figure 2**) [110, 111].

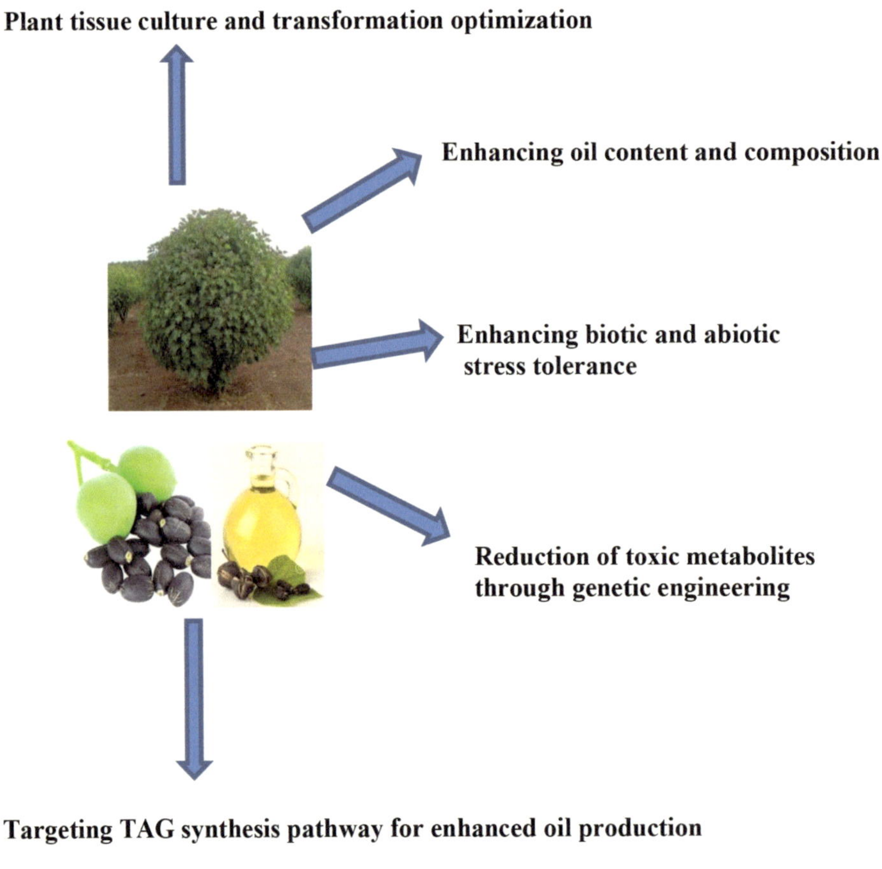

Figure 2.
Future perspectives of Jatropha for oil improvement.

15. Targeting enhanced oil production in *Jatropha curcus*

Various plant breeding strategies are employed in the last few decades to increase oil yield and quality, as well as resistance to biotic and abiotic challenges in edible and non-edible oil plants. Marker-aided selection, next-generation sequencing, "omics" technologies, and genetic engineering are some of the new biotechnological methods that have sped up the breeding process for such features in these plants. The use of omics technologies to identify and isolate important genes involved in lipid biosynthesis pathways, as well as their transfer to edible and non-edible oil plants is predicted to result in cost-effective oil production as a feedstock for biodiesel generation [112]. Biodiesel production from non-edible oil plants would be far more realistic if new varieties/hybrids of oil plants could be developed that contain more oil, are resistant to biotic and abiotic challenges, and do not contain harmful proteins. Through various breeding techniques, oil plants that produce edible and non-edible oils have increased these properties over recent decades. In the field of plant breeding, development, selection, target trait evaluation, multiplication, and distribution are the major objectives [113, 114]. Breeding targets for various crops have been rapidly accelerated by genetic and metabolic engineering techniques over the last few decades. Genetic engineering can be used to increase the amount of oil found in seeds of nonedible plants by engineering lipid biosynthesis pathways [115]. It is the simplest and most efficient way to increase oil yields in nonedible plants. Furthermore, the expression of genes encoding fatty acyl carrier thioesterase A (FatA), glycerol-3-phosphate dehydrogenase (GPD), and lysophosphatidyl-acyltransferases (LPAT) has enhanced the oil production pathway and therefore could be regarded as key genes to boost oil content in bioenergy plants [113, 116]. Engineering other genes involved in agronomical traits such as seed, fruit, and leaf size, plant growth and biomass, root architecture, and vegetative/reproductive transition, in addition to the genes involved in TAG biosynthesis.

Trait evaluation	Targeting gene	Reference
Inhibition of TAG degradation	Sugar-dependent protein 1 triacylglycerol lipase (SDP1) in *Jatropha*	[119]
Improving morphological and developmental traits	Auxin response factor 19 (JcARF19) in *Jatropha*	[120]
	Flowering locus T (JcFT)	[121]
Increased oleic acid content	Fatty acid desaturase (FAD2) in Jatropha	[122]
	Acyl-ACP thioesterase (FATB)	[123]
Reduction of toxins and inhibitors	Curcin precursor gene	[124]
	JcCASA: casbene synthase gene	[125]
Drought tolerance	PPAT; phosphopantetheine adenylyltransferase NF-YB: The subunit of the NF-Y transcription factor GSMT/DMT genes	[126]
Salinity tolerance	sbNhx1: encoding vacuolar Na+/H+ antiporter (NHX1)	[127]
Pest resistance	cry1Ab/1Ac: Bacillus thuringiensis dendotoxin	[128]
Disease resistance	Chitinase	[129]
	Hairpin dsRNA	[130]

Table 2.
Genetic manipulation strategies used to improve Jatropha.

Genetic engineering for oil content has a significant impact on the potential of bioenergy plants as a source of biodiesel production. Because seed size plays such an essential role in Jatropha oil yield, it has been prioritized as a breeding target to improve oil yields. In Jatropha, a candidate gene (CYP78A98) with the potential to increase seed size has just been discovered [117, 118]. Improvement of Jatropha through genetic engineering was listed in **Table 2**. The growing demand for biofuels has prompted plant scientists to develop plant feedstocks specifically for biodiesel production, using either traditional or modern breeding techniques to develop oil-seed varieties with higher oil content and optimal fatty acid composition. Biodiesel is a fuel made up of mono-alkyl esters of long-chain fatty acids derived from plant oils, with the majority of the fatty acids being triacylglycerols (TAGs) and short-chain alcohols (>95%). Waste vegetable oils and non-edible crude vegetable oils are another source of biodiesel that reduces its price. Jatropha, castor bean, cotton, Pongamia, tobacco, mahua, neem, and Camelina are currently used as non-edible oil yielding plants for second-generation biodiesel production [131]. Gene editing techniques like CRISPR can be used in precision breeding to improve yield, disease resistance, herbicide resistance, induce haploids, fix hybrid vigor, solve self-incompatibility, and help de novo domesticate oil crops. While it will likely be a long time before genome-edited oil crops become commercially available, we anticipate that regulatory constraints on them will gradually be eased in the near future [132].

Acknowledgements

The authors thank Department of Biotechnology, PSG College of Technology for providing infrastructural facilities and we thank Tamil Nadu Agricultural University for providing sesame seed varieties for our research purpose.

Conflict of interest

The authors declare no conflict of interest.

Author details

Muthulakshmi Chellamuthu[1], Kokiladevi Eswaran[2] and Selvi Subramanian[1*]

1 Department of Biotechnology, PSG College of Technology, Coimbatore, Tamil Nadu, India

2 Department of Plant Biotechnology, Center for Plant Molecular Biology and Biotechnology, Tamil Nadu Agricultural University, Coimbatore, Tamil Nadu, India

*Address all correspondence to: selvi.bio@psgtech.ac.in

References

[1] Datta A. GM crops: Dream to bring science to society. Agricultural Research. 2012;**1**(2):95-99

[2] Baulcombe D. Reaping benefits of crop research. Science. 2010;**327**(5967):761

[3] Ghosh S, Meli VS, Kumar A, Thakur A, Chakraborty N, Chakraborty S, et al. The N-glycan processing enzymes α-mannosidase and β-D-N-acetylhexosaminidase are involved in ripening-associated softening in the non-climacteric fruits of capsicum. Journal of Experimental Botany. 2011;**62**(2):571-582. DOI: 10.1093/jxb/erq289

[4] Kamthan A, Kamthan M, Azam M, Chakraborty N, Chakraborty S, Datta A. Expression of a fungal sterol desaturase improves tomato drought tolerance, pathogen resistance and nutritional quality. Scientific Reports. 2012;**2**(1): 951. DOI: 10.1038/srep00951

[5] James C. Global Status of Commercialized Biotech/GM Crops, 2011. Vol. 44. Ithaca, NY: ISAAA; 2011

[6] Hellenäs KE, Branzell C, Johnsson H, Slanina P. High levels of glycoalkaloids in the established Swedish potato variety Magnum Bonum. Journal of the Science of Food and Agriculture. 1995;**68**(2):249-255. DOI: 10.1002/jsfa.2740680217

[7] Ashri A. Sesame breeding. Plant Breeding Reviews. 1998;**16**:179-228

[8] Bedigian D. Evolution of sesame revisited: Domestication, diversity and prospects. Genetic Resources and Crop Evolution. 2003;**50**(7):779-787. DOI: 10.1023/A:1025029903549

[9] Wei X, Liu K, Zhang Y, Feng Q, Wang L, Zhao Y, et al. Genetic discovery for oil production and quality in sesame. Nature Communications. 2015;**6**:8609. DOI: 10.1038/ncomms9609

[10] Mondal N, Bhat KV, Srivastava PS. Variation in fatty acid composition in Indian germplasm of sesame. Journal of the American Oil Chemists Society. 2010;**87**:1263-1269. DOI: 10.1007/s11746-010-1615-9

[11] Pathak N, Rai AK, Kumari R, Thapa A, Bhat KV. Sesame crop: An underexploited oilseed holds tremendous potential for enhanced food value. Agricultural Sciences. 2014;**5**(06):519. DOI: 10.4236/as.2014.56054

[12] Gharby S, Harhar H, Bouzoubaa Z, Asdadi A, Yadini EI, Charrouf Z. Chemical characterization and oxidative stability of seeds and oil of sesame grown in Morocco. Journal of the Saudi society of Agricultural Science. 2017;**16**(2):105-111. DOI: 10.1016/j.jssas.2015.03.004

[13] Pathak N, Rai AK, Kumara R, Bhat KV. Value addition in sesame: A perspective on bioactive components for enhancing utility and profitability. Pharmacognosy Reviews. 2014;**8**(16): 147. DOI: 10.4103/0973-7847.134249

[14] Unal KM, Yalcin H. Proximate composition of Turkish sesame seeds and characterization of their oils. Grasas y Aceites. 2008;**59**(1):23-26

[15] Latif S, Anwar F. Aqueous enzymatic sesame oil and protein extraction. Food Chemistry. 2011;**125**(2):679-684. DOI: 10.1016/j.foodchem.2010.09.064

[16] Ashri A. Sesame (*Sesamum indicum* L.), genetic resources, chromosome engineering and crop improvement. In: Singh RJ, editor. Oilseed Crops. UK: CRC Press; 2007. pp. 231-289

[17] Khan MA, Mirza MY, Akmal M, Ali N, Khan I. Genetic parameters and

their implications for yield improvement in sesame. Sarhad Journal of Agriculture. 2007;**23**(3):623

[18] Sarwar G, Hussain J. Selection criteria In M3 and M4 populations of sesame (*Sesamum Indicum* L.). Journal of Agricultural Research. 2010;**48**(1):1-13

[19] Yadav LN, Tripathi MK, Sikarwar RS, Mishra AK. Heterosis in sesame. Sesame and Safflower Newsletter. 2005;**20**

[20] Hoballah AA. Selection and agronomic evaluation of induced mutant lines of sesame. 2001:137-149

[21] Arslan C, Uzun B, Ülger S, İlhan CM. Determination of oil content and fatty acid composition of sesame mutants suited for intensive management conditions. Journal of the American Oil Chemist Society. 2007;**84**:917-920. DOI: 10.1007/s11746-007-1125-6

[22] Mary RJ, Jayabalan N. Influence of growth regulators on somatic embryogenesis in sesame. Plant cell, Tissue and Organ Culture. 1997;**49**(1):67-70. DOI: 10.1023/A:1005808701747

[23] Shashidhara N, Ravikumar H, Ashoka N, Santosh DT, Pawar P, Lokesha R, et al. Callus induction and sub-culturing in sesame (Sesamum indicum L.): A basic strategy. International Journal of Agriculture, Environment and Biotechnology. 2011;**4**(2):153-156

[24] Zimik M, Arumugam N. Induction of shoot regeneration in cotyledon explants of the oilseed crop *Sesamum indicum* L. Journal of Genetic Engineering and Biotechnology. 2017;**15**(2):303-308. DOI: 10.1016/j.jgeb.2017.07.006

[25] Yadav M, Chaudhary D, Sainger M, Jaiwal PK. Agrobacterium tumefaciens-mediated genetic transformation of sesame (*Sesamum indicum* L.). Plant Cell, Tissue and Organ Culture (PCTOC). 2010;**103**(3):377-386. DOI: 10.1007/s11240-010-9791-8

[26] Parveen S, Mazumder M, Bhattacharya A, Mukhopadhyay S, Saha U, Mukherjee A, et al. Identification of anther-specific genes from sesame and functional assessment of the upstream region of a Tapetum-specific β-1, 3-glucanase gene. Plant Molecular Biology Reporter. 2018; **36**(2):149-161. DOI: 10.1007/s11105-017-1054-y

[27] Chowdhury S, Basu A, Kundu S. A new high-frequency agrobacterium-mediated transformation technique for *Sesamum indicum* L. using de-embryonated cotyledon as explant. Protoplasma. 2014;**251**(5):1175-1190

[28] Khan AW, Garg V, Roorkiwal M, Golicz AA, Edwards D, Varshney RK. Super-pangenome by integrating the wild side of a species for accelerated crop improvement. Trends in Plant Science. 2020;**25**(2):148-158. DOI: 10.1016/j.tplants.2019.10.012

[29] Zhang H, Miao H, Wang LQL, Liu H, Wang Q, Yue M. Genome sequencing of the important oilseed crop *Sesamum indicum* L. Genome Biology. 2013;**14**:401. DOI: 10.1186/gb-2013-14-1-401

[30] Bhunia RK, Chakraborty A, Kaur R, Maiti MK, Sen SK. Enhancement of α-linolenic acid content in transgenic tobacco seeds by targeting a plastidial ω-3 fatty acid desaturase (fad7) gene of Sesamum indicum to ER. Plant cell reports. 2016;**35**(1):213-226. DOI: 10.1007/s00299-015-1880-z

[31] Chellamuthu M, Kumaresan K, Subramanian S, Muthumanickam H. Functional analysis of sesame diacylglycerol acyltransferase and

phospholipid: Diacylglycerol acyltransferase genes using in silico and in vitro approaches. Plant Molecular Biology Reporter. 2019;**37**(3):146-156. DOI: 10.1007/s11105-019-01144-7

[32] Muthulakshmi C, Sivaranjani R, Subramanian S. Modification of sesame (*Sesamum indicum* L) for triacylglycerol accumulation in plant biomass for biofuel applications. Biotechnology Reports. 2021;**32**:e00668. DOI: 10.1016/j. btre.2021.e00668

[33] Pasupuleti J, Nigam SN, Pandey MK, Nagesh P, Varshney RK. Groundnut improvement: Use of genetic and genomic tools. Frontiers in Plant Science. 2013;**4**:23. DOI: 10.3389/fpls.2013.00023

[34] Chu Y, Deng XY, Faustinelli P, Ozias-Akins P. Bcl-xL transformed peanut (*Arachis hypogaea* L.) exhibits paraquat tolerance. Plant Cell Reports. 2008;**27**(1):85-92. DOI: 10.1007/s00299-007-0444-2

[35] Asif MA, Zafar Y, Iqbal J, Iqbal MM, Rashid U, Ali GM, et al. Enhanced expression of AtNHX1, in transgenic groundnut (*Arachis hypogaea* L.) improves salt and drought tolerence. Molecular Biotechnology. 2011;**49**(3):250-256. DOI: 10.1007/s12033-011-9399-1

[36] Manjulatha M, Sreevathsa R, Kumar AM, Sudhakar C, Prasad TG, Tuteja N, et al. Overexpression of a pea DNA helicase (PDH45) in peanut (*Arachis hypogaea* L.) confers improvement of cellular level tolerance and productivity under drought stress. Molecular Biotechnology. 2014;**56**(2):111-125. DOI: 10.1007/s12033-013-9687-z

[37] Pandurangaiah M, Rao GL, Sudhakarbabu O, Nareshkumar A, Kiranmai K, Lokesh U, et al. Overexpression of horsegram (*Macrotyloma uniflorum* Lam. Verdc.) NAC transcriptional factor (mu NAC4) in groundnut confers enhanced drought tolerance. Molecular Biotechnology. 2014;**56**(8):758-769. DOI: 10.1007/s12033-014-9754-0

[38] Patil M, Ramu SV, Jathish P, Sreevathsa R, Reddy PC, Prasad TG, et al. Overexpression of AtNAC2 (ANAC092) in groundnut (*Arachis hypogaea* L.) improves abiotic stress tolerance. Plant Biotechnology Reports. 2014;**8**(2):161-169. DOI: 10.1007/s11816-013-0305-0

[39] Nageswara Rao RC, Nigam SN. Genetic options for drought management in groundnut. 2003:123-141

[40] Ratnakumar P, Vadez V, Nigam SN, Krishnamurthy L. Assessment of transpiration efficiency in peanut (*Arachis hypogaea* L.) under drought using a lysimetric system. Plant Biology. 2009;**11**:124-130. DOI: 10.1111/j.1438-8677.2009.00260.x

[41] Nageswara Rao RC, Talwar HS, Wright GC. Rapid assessment of specific leaf area and leaf nitrogen in peanut (*Arachis hypogaea* L.) using a chlorophyll meter. Journal of Agronomy and Crop Science. 2001;**186**(3):175-182. DOI: 10.1046/j.1439-037X.2001.00472.x

[42] Sheshshayee MS, Bindumadhava H, Shankar AG, Prasad TG, Udayakumar M. Breeding strategies to exploit water use efficiency for crop improvement. Journal of Plant Biology-New Delhi. 2003;**30**(2):253-268

[43] Craufurd PQ, Prasad PV, Summerfield RJ. Dry matter production and rate of change of harvest index at high temperature in peanut. Crop Science. 2002;**42**(1):146-151. DOI: 10.2135/cropsci2002.1460

[44] Craufurd PQ, Prasad PV, Kakani VG, Wheeler TR, Nigam SN. Heat tolerance in groundnut. Field Crops Research. 2003;**80**(1):63-77

[45] Vadez V, Srivastava N, Krishnamurthy L, Aruna R, Nigam SN. Standardization of a protocol to screen for salinity tolerance in groundnut. International Arachis Newsletter. 2005;**25**:42-47

[46] Singh AL, Hariprassana K, Solanki RM. Screening and selection of groundnut genotypes for tolerance of soil salinity. 2008;**1**(3):69-77

[47] Yang Q, Jin H. The effect of aluminum on N, P and Ca uptake in peanut varieties. Chinese Journal of Oil Crop Sciences. 2000;**22**(2):68-73

[48] Pradap S, Ramarao G, Reddy KB. Response of groundnut genotypes to aluminium toxicity in solution culture. 2002;7(4):396-400

[49] Qin H, Gu Q, Zhang J, Sun L, Kuppu S, Zhang Y, et al. Regulated expression of an isopentenyltransferase gene (IPT) in peanut significantly improves drought tolerance and increases yield under field conditions. Plant and Cell Physiology. 2011;**52**(11):1904-1914. DOI: 10.1093/pcp/pcr125

[50] Bhatnagar-Mathur P, Devi MJ, Reddy DS, Lavanya M, Vadez V, Serraj R, et al. Stress-inducible expression of At DREB1A in transgenic peanut (*Arachis hypogaea* L.) increases transpiration efficiency under water-limiting conditions. Plant Cell Reports. 2007;**26**(12):2071-2082. DOI: 10.1007/s00299-007-0406-8

[51] Bhatnagar-Mathur P, Devi MJ, Vadez V, Sharma KK. Differential antioxidative responses in transgenic peanut bear no relationship to their superior transpiration efficiency under drought stress. Journal of Plant Physiology. 2009;**166**(11):1207-1217. DOI: 10.1016/j.jplph.2009.01.001

[52] Bhatnagar-Mathur P, Rao JS, Vadez V, Dumbala SR, Rathore A, Yamaguchi-Shinozaki K, et al. Transgenic peanut overexpressing the DREB1A transcription factor has higher yields under drought stress. Molecular Breeding. 2014;**33**(2):327-340. DOI: 10.1007/s11032-013-9952-7

[53] Singh N, Mishra A, Jha B. Ectopic over-expression of peroxisomal ascorbate peroxidase (SbpAPX) gene confers salt stress tolerance in transgenic peanut (Arachis hypogaea). Gene. 2014;**547**(1):119-125. DOI: 10.1016/j.gene.2014.06.037

[54] Bhauso TD, Radhakrishnan T, Kumar A, Mishra GP, Dobaria JR, Patel K, et al. Overexpression of bacterial mtlD gene in peanut improves drought tolerance through accumulation of mannitol. The Scientific World Journal. 2014;**125967**. DOI: 10.1155/2014/125967

[55] Devi MJ, Sinclair TR, Vadez V. Genotypic variability among peanut (*Arachis hypogea* L.) in sensitivity of nitrogen fixation to soil drying. Plant and Soil. 2010;**330**(1):139-148. DOI: 10.1007/s11104-009-0185-9

[56] Qiao LX, Ding X, Wang HC, Sui JM, Wang JS. Characterization of the β-1, 3-glucanase gene in peanut (*Arachis hypogaea* L.) by cloning and genetic transformation. Genetics and Molecular Research. 2014;**13**:1893-1904. DOI: 10.4238/2014.March.17.17

[57] Kumar D, Kirti PB. Transcriptomic and proteomic analyses of resistant host responses in Arachis diogoi challenged with late leaf spot pathogen, *Phaeoisariopsis personata*. PLoS one. 2015;**10**(2):e0117559. DOI: 10.1371/journal.pone.0117559

[58] Prasad K, Bhatnagar-Mathur P, Waliyar F, Sharma KK. Overexpression of a chitinase gene in transgenic peanut confers enhanced resistance to major soil borne and foliar fungal pathogens. Journal of Plant Biochemistry and

Biotechnology. 2013;**22**(2):222-233. DOI: 10.1007/s13562-012-0155-9

[59] Vasavirama K, Kirti PB. Increased resistance to late leaf spot disease in transgenic peanut using a combination of PR genes. Functional & Integrative Genomics. 2012;**12**(4):625-634. DOI: 10.1007/s10142-012-0298-8

[60] Sundaresha S, Kumar AM, Rohini S, Math SA, Keshamma E, Chandrashekar SC, et al. Enhanced protection against two major fungal pathogens of groundnut, *Cercospora arachidicola* and *Aspergillus flavus* in transgenic groundnut over-expressing a tobacco β 1-3 glucanase. European Journal of Plant Pathology. 2010; **126**(4):497-508. DOI: 10.1007/s10658-009-9556-6

[61] Mehta R, Radhakrishnan T, Kumar A, Yadav R, Dobaria JR, Thirumalaisamy PP, et al. Coat protein-mediated transgenic resistance of peanut (*Arachis hypogaea* L.) to peanut stem necrosis disease through agrobacterium-mediated genetic transformation. Indian Journal of Virology. 2013;**24**(2):205-213. DOI: 10.1007/s13337-013-0157-9

[62] Rao SC, Bhatnagar-Mathur P, Kumar PL, Reddy AS, Sharma KK. Pathogen-derived resistance using a viral nucleocapsid gene confers only partial non-durable protection in peanut against peanut bud necrosis virus. Archives of Virology. 2013;**158**(1):133-143. DOI: 10.1007/s00705-012-1483-8

[63] Higgins CM, Hall RM, Mitter N, Cruickshank A, Dietzgen RG. Peanut stripe potyvirus resistance in peanut (*Arachis hypogaea* L.) plants carrying viral coat protein gene sequences. Transgenic Research. 2004;**13**(1):59-67. DOI: 10.1023/B:TRAG.0000017166.29458.74

[64] Magbanua ZV, Wilde HD, Roberts JK, Chowdhury K, Abad J, Moyer JW, et al. Field resistance to tomato spotted wilt virus in transgenic peanut (*Arachis hypogaea* L.) expressing an antisense nucleocapsid gene sequence. Molecular Breeding. 2000;**6**(2):227-236. DOI: 10.1023/A:1009649408157

[65] Keshavareddy G, Rohini S, Ramu SV, Sundaresha S, Kumar AR, Kumar PA, et al. Transgenics in groundnut (*Arachis hypogaea* L.) expressing cry1AcF gene for resistance to *Spodoptera litura* (F.). Physiology and Molecular Biology of Plants. 2013;**19**(3):343-352. DOI: 10.1007/s12298-013-0182-6

[66] Geng L, Niu L, Gresshoff PM, Shu C, Song F, Huang D, et al. Efficient production of *Agrobacterium rhizogenes*-transformed roots and composite plants in peanut (*Arachis hypogaea* L.). Plant Cell, Tissue and Organ Culture (PCTOC). 2012;**109**(3):491-500. DOI: 10.1007/s11240-012-0113-1

[67] Livingstone DM, Hampton JL, Phipps PM, Grabau EA. Enhancing resistance to *Sclerotinia minor* in peanut by expressing a barley oxalate oxidase gene. Plant Physiology. 2005;**137**(4): 1354-1362. DOI: 10.1104/pp.104.057232

[68] Partridge-Telenko DE, Hu J, Livingstone DM, Shew BB, Phipps PM, Grabau EA. Sclerotinia blight resistance in Virginia-type peanut transformed with a barley oxalate oxidase gene. Phytopathology. 2011;**101**(7):786-793. DOI: 10.1094/PHYTO-10-10-0266

[69] Tiwari S, Mishra DK, Chandrasekhar K, Singh PK, Tuli R. Expression of δ-endotoxin Cry1EC from an inducible promoter confers insect protection in peanut (*Arachis hypogaea* L.) plants. Pest Management Science. 2011;**67**(2):137-145. DOI: 10.1002/ps.2041

[70] Singsit C, Adang MJ, Lynch RE, Anderson WF, Wang A, Cardineau G,

et al. Expression of a *Bacillus thuringiensis* cryIA (c) gene in transgenic peanut plants and its efficacy against lesser cornstalk borer. Transgenic Research. 1997;**6**(2):169-176. DOI: 10.1023/A:1018481805928

[71] Li XY, Liu X, Yao Y, Li YH, Liu S, He CY, et al. Overexpression of *Arachis hypogaea* AREB1 gene enhances drought tolerance by modulating ROS scavenging and maintaining endogenous ABA content. International Journal of Molecular Sciences. 2013;**14**(6):12827-12842. DOI: 10.3390/ijms140612827

[72] Yang CY, Chen SY, Duan GC. Transgenic peanut (*Arachis hypogaea* L.) expressing the urease subunit B gene of *Helicobacter pylori*. Current Microbiology. 2011;**63**(4):387-391. DOI: 10.1007/s00284-011-9991-4

[73] Athmaram TN, Bali G, Devaiah KM. Integration and expression of Bluetongue VP2 gene in somatic embryos of peanut through particle bombardment method. Vaccine. 2006;**24**(15):2994-3000. DOI: 10.1016/j.vaccine.2005.11.019

[74] Dodo HW, Konan KN, Chen FC, Egnin M, Viquez OM. Alleviating peanut allergy using genetic engineering: The silencing of the immunodominant allergen Ara h 2 leads to its significant reduction and a decrease in peanut allergenicity. Plant Biotechnology Journal. 2008;**6**(2):135-145. DOI: 10.1111/j.1467-7652.2007.00292.x

[75] Yuan M, Zhu J, Gong L, He L, Lee C, Han S, et al. Mutagenesis of FAD2 genes in peanut with CRISPR/Cas9 based gene editing. BMC Biotechnology. 2019;**19**(1):1-7. DOI: 10.1186/s12896-019-0516-8

[76] Wen S, Liu H, Li X, Chen X, Hong Y, Li H, et al. TALEN-mediated targeted mutagenesis of fatty acid desaturase 2 (FAD2) in peanut (*Arachis hypogaea* L.) promotes the accumulation of oleic acid. Plant Molecular Biology. 2018;**97**(1):177-185. DOI: 10.1007/s11103-018-0731-z

[77] Cong L, Ran FA, Cox D, Lin S, Barretto R, Habib N, et al. Multiplex genome engineering using CRISPR/Cas systems. Science. 2013;**339**(6121):819-823. DOI: 10.1126/science.1231143

[78] Nekrasov V, Staskawicz B, Weigel D, Jones JD, Kamoun S. Targeted mutagenesis in the model plant *Nicotiana benthamiana* using Cas9 RNA-guided endonuclease. Nature Biotechnology. 2013;**31**(8):691-693. DOI: 10.1038/nbt.2655

[79] Sugano SS, Osakabe K, Osakabe Y. Crop breeding using CRISPR/Cas9. InCrop Improvement through Microbial Biotechnology. 2018:451-464. DOI: 10.1016/B978-0-444-63987-5.00023-2

[80] Yin K, Gao C, Qiu JL. Progress and prospects in plant genome editing. Nature Plants. 2017;**3**(8):1-6. DOI: 10.1038/nplants.2017.107

[81] Borrelli VM, Brambilla V, Rogowsky P, Marocco A, Lanubile A. The enhancement of plant disease resistance using CRISPR/Cas9 technology, 10.3389/fpls.2018.01245. Frontiers in Plant Science. 2018;**9**:1245

[82] Shu H, Luo Z, Peng Z, et al. The application of CRISPR/Cas9 in hairy roots to explore the functions of *AhNFR1* and *AhNFR5* genes during peanut nodulation. BMC Plant Biology. 2020;**20**:417. DOI: 10.1186/s12870-020-02614-x

[83] Dajue L, Mundel HH. Safflower. *Carthamus tinctorius* (L.) Promoting the Conservation and Use of Underutilized and Neglected Crops. Rome, Italy: Institute of Plant Genetics and Crop Plant Research, Gaterslaben/International Plant Genetic Resources Institute; 1996. p. 83

[84] Zhang LP. Safflower: A versatile plant. In: Corleto A, Mundel HH, editors. Proceedings of the IVth International Safflower Conference. Bari, Italy: Adriatica Editrice; 1997. pp. 329-331

[85] Nykiforuk CL, Shen Y, Murray EW, Boothe JG, Busseuil D, Rheaume E, et al. Expression and recovery of biologically active recombinant apolipoprotein AIMilano from transgenic safflower (*Carthamus tinctorius*) seeds. Plant Biotechnology Journal. 2011;**9**(2):250-263. DOI: 10.1111/j.1467-7652.2010.00546.x

[86] Markley N, Nykiforuk C, Boothe J, Moloney M. Producing proteins using oilbody-oleosin technology. BioPharm International. 2006;**9**:6-12

[87] Ying M, Dyer WE, Bergman JW. Agrobacterium tumefaciens-mediated transformation of safflower (*Carthamus tinctorius* L.) cv. 'Centennial'. Plant Cell Reports. 1992;**11**(11):581-585. DOI: 10.1007/BF00233097

[88] Sankararao K, Rohini VK. Gene transfer into Indian cultivars of safflower (*Carthamus tinctorius* L.) using agrobacterium tumefaciens. Plant Biotechnology. 1999;**16**(3):201-206. DOI: 10.5511/plantbiotechnology.16.201

[89] Mündel HH, Blackshaw RE, Byers JR, Huang HC, Johnson DL, Keon R, et al. Safflower Production on the Canadian Prairies: Revisited in 2004. Agriculture and Agri-Food Canada; 2004

[90] Velasco L, Pérez-Vich B, Fernández-Martínez JM. Identification and genetic characterization of a safflower mutant with a modified tocopherol profile. Plant Breeding. 2005;**124**(5):459-463. DOI: 10.1111/j.1439-0523.2005.01150.x

[91] Mihaela P, Josef R, Monica N, Rudolf Z. Perspectives of safflower oil as biodiesel source for South Eastern Europe (comparative study: Safflower, soybean and rapeseed). Fuel. 2013;**111**:114-119. DOI: 10.1016/j.fuel.2013.04.012

[92] Sehgal D, Raina SN. Carthamus. In: Kole C, editor. Wild Crop Relatives: Genomic and Breeding Resources. Berlin, Heidelberg: Springer; 2011. pp. 63-95

[93] Golkar P. Breeding improvements in safflower ('*Carthamus tinctorius*' L.): A review. Australian Journal of Crop Science. 2014;**8**(7):1079-1085

[94] Singh V, Ranaware AM, Nimbkar N. Breeding for Fusarium wilt resistance in safflower. In: Paper Presented at: Safflower: Unexploited Potential and World Adaptability 7th International Safflower Conference. NSW, Australia: Wagga Wagga; 2008

[95] Meena HP, Dudhe MY, Mukta N, Anjani K. Heterosis breeding in safflower: Present status and future prospects under Indian scenario. Journal of Oilseeds Research. 2012;**29**:164-167

[96] Hamdan YA, Velasco L, Pérez-Vich B. Development of SCAR markers linked to male sterility and very high linoleic acid content in safflower. Molecular Breeding. 2008;**22**(3):385-393. DOI: 10.1007/s11032-008-9183-5

[97] Rapson S, Wu M, Okada S, Das A, Shrestha P, Zhou XR, et al. A case study on the genetic origin of the high oleic acid trait through FAD2-1 DNA sequence variation in safflower (*Carthamus tinctorius* L.). Frontiers in Plant Science. 2015;**6**:691. DOI: 10.3389/fpls.2015.00691

[98] Cao S, Zhu QH, Shen W, Jiao X, Zhao X, Wang MB, et al. Comparative profiling of miRNA expression in developing seeds of high linoleic and high oleic safflower (*Carthamus tinctorius* L.) plants. Frontiers in Plant

Science. 2013;**4**:489. DOI: 10.3389/fpls.2013.00489

[99] Liu Q, Cao S, Zhou XR, Wood C, Green A, Singh S. Nonsense-mediated mRNA degradation of CtFAD2-1 and development of a perfect molecular marker for olol mutation in high oleic safflower (*Carthamus tinctorius* L.). Theoretical and Applied Genetics. 2013;**126**(9):2219-2231. DOI: 10.1007/s00122-013-2129-2

[100] Vanhercke T, Wood CC, Stymne S, Singh SP, Green AG. Metabolic engineering of plant oils and waxes for use as industrial feedstocks. Plant Biotechnology Journal. 2013;**11**(2):197-210. DOI: 10.1111/pbi.12023

[101] Wood CC, Okada S, Taylor MC, Menon A, Mathew A, Cullerne D, et al. Seed-specific RNAi in safflower generates a superhigh oleic oil with extended oxidative stability. Plant Biotechnology Journal. 2018;**16**(10):1788-1796. DOI: 10.1111/pbi.12915

[102] Rani A, Panwar A, Sathe M, Chandrashekhara KA, Kush A. Biofortification of safflower: An oil seed crop engineered for ALA-targeting better sustainability and plant based omega-3 fatty acids. Transgenic Research. 2018;**27**(3):253-263. DOI: 10.1007/s11248-018-0070-5

[103] Divakara BN, Upadhyaya HD, Wani SP, Gowda CL. Biology and genetic improvement of *Jatropha curcas* L.: A review. Applied Energy. 2010;**87**(3):732-742. DOI: 10.1016/j.apenergy.2009.07.013

[104] Pandey VC, Singh K, Singh JS, Kumar A, Singh B, Singh RP. Jatropha curcas: A potential biofuel plant for sustainable environmental development. Renewable and Sustainable Energy Reviews. 2012;**16**(5):2870-2883. DOI: 10.1016/j.rser.2012.02.004

[105] Ilham Z, Saka S. Two-step supercritical dimethyl carbonate method for biodiesel production from *Jatropha curcas* oil. Bioresource Technology. 2010;**101**(8):2735-2740. DOI: 10.1016/j.biortech.2009.10.053

[106] Berchmans HJ, Hirata S. Biodiesel production from crude *Jatropha curcas* L. seed oil with a high content of free fatty acids. Bioresource Technology. 2008;**99**(6):1716-1721. DOI: 10.1016/j.biortech.2007.03.051

[107] Chhetri AB, Tango MS, Budge SM, Watts KC, Islam MR. Non-edible plant oils as new sources for biodiesel production. International Journal of Molecular Sciences. 2008;**9**(2):169-180. DOI: 10.3390/ijms9020169

[108] Achten WM, Maes WH, Aerts R, Verchot L, Trabucco A, Mathijs E, et al. Jatropha: From global hype to local opportunity. Journal of Arid Environments. 2010;**74**(1):164-165. DOI: 10.1016/j.jaridenv.2009.08.010

[109] Dyer JC, Stringer LC, Dougill AJ. *Jatropha curcas*: Sowing local seeds of success in Malawi?: In response to Achten et al.(2010). Journal of Arid Environments. 2012;**79**:107-110. DOI: 10.1016/j.jaridenv.2011.12.004

[110] Ravindranath NH, Lakshmi CS, Manuvie R, Balachandra P. Biofuel production and implications for land use, food production and environment in India. Energy Policy. 2011;**39**(10):5737-5745. DOI: 10.1016/j.enpol.2010.07.044

[111] Joshi M, Mishra A, Jha B. Efficient genetic transformation of *Jatropha curcas* L. by microprojectile bombardment using embryo axes. Industrial Crops and Products. 2011;**33**(1):67-77. DOI: 10.1016/j.indcrop.2010.09.002

[112] Grover A, Patade VY, Kumari M, Gupta SM, Arif M, Ahmed Z. Omics

approaches in biofuel production for a green environment. OMICS: Applications in Biomedical, Agricultural and Environment Sciences. 2013;**26**:623-636. DOI: 10.1201/b14289-29

[113] Singh AS, Kumari S, Modi AR, Gajera BB, Narayanan S, Kumar N. Role of conventional and biotechnological approaches in genetic improvement of castor (*Ricinus communis* L.). Industrial Crops and Products. 2015;**74**:55-62. DOI: 10.1016/j.indcrop.2015.05.001

[114] Moniruzzaman M, Yaakob Z, Khatun R. Biotechnology for Jatropha improvement: A worthy exploration. Renewable and Sustainable Energy Reviews. 2016;**54**:1262-1277. DOI: 10.1016/j.rser.2015.10.074

[115] Hegde K, Chandra N, Sarma SJ, Brar SK, Veeranki VD. Genetic engineering strategies for enhanced biodiesel production. Molecular Biotechnology. 2015;**57**(7):606-624. DOI: 10.1007/s12033-015-9869-y

[116] Liu Q, Guo Q, Akbar S, Zhi Y, El Tahchy A, Mitchell M, et al. Genetic enhancement of oil content in potato tuber (*Solanum tuberosum* L.) through an integrated metabolic engineering strategy. Plant Biotechnology Journal. 2017;**15**(1):56-67. DOI: 10.1111/pbi.12590

[117] Akashi K, Nanasato Y. Recent progress in the genetic engineering of biofuel crops. Biofuels: Greenhouse Gas Mitigation and Global Warming. 2018:327-339. DOI: 10.1007/978-81-322-3763-1_18

[118] Tian Y, Zhang M, Hu X, Wang L, Dai J, Xu Y, et al. Over-expression of CYP78A98, a cytochrome P450 gene from *Jatropha curcas* L., increases seed size of transgenic tobacco. Electronic Journal of Biotechnology. 2016;**19**:15-22. DOI: 10.1016/j.ejbt.2015.11.001

[119] Chua NH, Kim MJ, Yang SW, inventors; Temasek Life Sciences Laboratory Ltd, assignee. Gene silencing of sugar-dependent 1 in *Jatropha curcas*. United States patent US 9,650,644. 2017

[120] Sun Y, Wang C, Wang N, Jiang X, Mao H, Zhu C, et al. Manipulation of auxin response factor 19 affects seed size in the woody perennial *Jatropha curcas*. Scientific Reports. 2017;**7**(1):1-2. DOI: 10.1038/srep40844

[121] Li C, Luo L, Fu Q, Niu L, Xu ZF. Isolation and functional characterization of JcFT, a FLOWERING LOCUS T (FT) homologous gene from the biofuel plant Jatropha curcas. BMC Plant Biology. 2014;**14**(1):1-2. DOI: 10.1186/1471-2229-14-125

[122] Qu J, Mao HZ, Chen W, Gao SQ, Bai YN, Sun YW, et al. Development of marker-free transgenic Jatropha plants with increased levels of seed oleic acid. Biotechnology for Biofuels. 2012;**5**(1):1-1. DOI: 10.1186/1754-6834-5-10

[123] Buhr T, Sato S, Ebrahim F, Xing A, Zhou Y, Mathiesen M, et al. Ribozyme termination of RNA transcripts down-regulate seed fatty acid genes in transgenic soybean. The Plant Journal. 2002;**30**(2):155-163. DOI: 10.1046/j.1365-313X.2002.01283.x

[124] Patade VY, Khatri D, Kumar K, Grover A, Kumari M, Gupta SM, et al. RNAi mediated curcin precursor gene silencing in Jatropha (*Jatropha curcas* L.). Molecular Biology Reports. 2014;**41**(7):4305-4312. DOI: 10.1007/s11033-014-3301-8

[125] Li Y, Huang Y, Du W, Dai L, Liu D. Combined phospholipase and lipase catalysis for biodiesel production from phospholipids-containing oil. Biotechnology and Bioprocess Engineering. 2015;**20**(5):965-970. DOI: 10.1007/s12257-015-0250-6

[126] Tsuchimoto S, Cartagena J, Khemkladngoen N, Singkaravanit S,

Kohinata T, Wada N, et al. Development of transgenic plants in jatropha with drought tolerance. Plant Biotechnology. 2012:12-0406. DOI: 10.5511/plantbiotechnology.12.0406d

[127] Jha B, Mishra A, Jha A, Joshi M. Developing transgenic Jatropha using the SbNHX1 gene from an extreme halophyte for cultivation in saline wasteland. PLoS One. 2013;**8**(8):e71136. DOI: 10.1371/journal.pone.0071136

[128] Gu K, Mao H, Yin Z. Production of marker-free transgenic *Jatropha curcas* expressing hybrid bacillus thuringiensis δ-endotoxin Cry1Ab/1Ac for resistance to larvae of tortrix moth (*Archips micaceanus*). Biotechnology for Biofuels. 2014;**7**(1):1-9. DOI: 10.1186/1754-6834-7-68

[129] Franco MC, Gomes KA, de Carvalho Filho MM, Harakava R, Carels N, Siqueira WJ, et al. Agrobacterium-mediated transformation of *Jatropha curcas* leaf explants with a fungal chitinase gene. African Journal of Biotechnology. 2016;**15**(37):2006-2016. DOI: 10.5897/AJB2016.15458

[130] Ye J, Qu J, Mao HZ, Ma ZG, Rahman NE, Bai C, et al. Engineering geminivirus resistance in *Jatropha curcus*. Biotechnology for Biofuels. 2014;**7**(1):1-1. DOI: 10.1186/s13068-014-0149-z

[131] He J, Zhang K, Tang M, Zhou W, Chen L, Chen Z, et al. CRISPR-based genome editing technology and its applications in oil crops. Oil Crop Science. 2021;**6**:105-113. DOI: 10.1016/j.ocsci.2021.07.002

[132] Miladinovic D, Antunes D, Yildirim K, et al. Targeted plant improvement through genome editing: From laboratory to field. Plant Cell Reports. 2021;**40**:935-951. DOI: 10.1007/s00299-020-02655-4

Chapter 2

Genetic Modification and Application in Cassava, Sweetpotato and Yams

Prince Emmanuel Norman, Daniel K. Dzidzienyo, Kumba Yannah Karim and Aloysius A. Beah

Abstract

Cassava (*Manihot esculenta* Crantz), sweetpotato (*Ipomoea batatas*) and yams (Dioscorea spp.) are important root and tuber crops grown for food, feed and various industrial applications. However, their genetic gain potentials are limited by breeding and genetic bottlenecks for improvement of many desired traits. This book chapter covers the applications and potential benefits of genetic modification in breeding selected outcrossing root and tuber crops. It assesses how improvement of selected root and tuber crops through genetic modification overcomes both the high heterozygosity and serious trait separation that occurs in conventional breeding, and contributes to timely achievement of improved target traits. It also assesses the ways genetic modification improves genetic gain in the root and tuber breeding programs, conclusions and perspectives. Conscious use of complementary techniques such as genetic modification in the root and tuber breeding programs can increase the selection gain by reducing the long breeding cycle and cost, as well as reliable exploitation of the heritable variation in the desired direction.

Keywords: application, genetic modification, genetic gain, transgenic plants, roots and tubers

1. Introduction

Root and tuber crops including cassava (*Manihot esculenta* Crantz), sweetpotato (*Ipomoea batatas*) and yams (Dioscorea spp.) are important crops with increasing food, feed and industrial applications in Sub-Saharan Africa and many other regions of the world [1–3]. These crops possess great potential to contribute to food, nutrition and income security of many livelihoods worldwide, but this potential is to be fully exploited. Variety development through breeding is among the activities targeted at unlocking the potential of these crops for food, feed and industrial applications [3].

Despite their importance, conventional breeding of root and tuber crops is limited by many challenges and heavily depends on the traditional techniques for exploitation of the existing variation. These challenges include high degree of genetic heterozygosity, genetic overloading, serious separation of progeny, few flowers, Irregularity in flowering time and flowering intensity, low pollen fertility, self-incompatibility, cross incompatibility, polyploidy, and low fruit set rate [3–6].

IntechOpen

Genetic modification technologies are among many advances made to traditional breeding practices in plants, animals, and microbes to increase productivity and enhance food quality. Plant genetic modification is the oldest technique utilized in simple selection, where plants that exhibit desired characteristics are selected for continued propagation [7]. The advent of modern technology and various molecular analytical tools has improved upon simple selection for detection of elite plants expressing desired traits. Genetic modification is an important alternative and complementary technique for the genetic improvement of crops including roots and tubers. It is a powerful tool that can be used to introduce a number of genes with important agronomic traits, such as disease resistance, insect resistance, and high yield and quality. Genetic modification shows great potential for the genetic improvement of crops including roots and tubers and can compensate for the limitations of conventional breeding. The application of transgenic methods to cassava, sweetpotato and yam improvement programs is particularly important due to the difficulties associated with conventional breeding of these crops. However, an efficient plant regeneration system is imperative to achieve successful transformation [8].

Since the advent of genetic modification, rapid progress has been noted for cassava, sweetpotato and yam breeding programs through various of international non-profit organizations and scientists from developed countries. For instance, HarvestPlus and BioCassava Plus, have made remarkable achievements by transforming conventional breeding into molecular breeding [9, 10]. This book chapter focuses on genetic modification in selected root and tuber crops, applications, potential and future prospects for the genetic improvement of these economically important crops.

1.1 Concepts of genetic and nongenetic modifications

Genetic modification or transformation is the directed desirable transfer of gene or insertion of DNA from one organism to another along with the subsequent stable integration and expression of a foreign gene in the genome [7]. It also refers to the targeted manipulation of genetic material, and nontargeted, nontransgenic methods such as chemical mutagenesis and breeding applied to alter the genetic composition of plants, animals, and microorganisms. Genetic engineering refers to recombinant deoxyribonucleic acid (rDNA) methods that permit a gene from any species to be inserted and subsequently expressed in a food crop or other food product [7]. Although the process involving rDNA technology is not inherently hazardous, the products of rDNA technology may only be hazardous if inserted genes result in the production of hazardous substances.

Nongenetic engineering techniques of genetic modification such as embryo rescue involves placing of plant or animal embryos produced from interspecies gene transfer, or crossing, in a tissue culture environment to complete development [7]. Somatic hybridization of nongenetic engineering technique involves removal of the cell walls of a plant, forceful hybridization of cells and induction of mutagenesis. Irradiation or chemical mutagenesis is useful for the induction of random mutations in DNA [7]. The development of genetic modification approaches has enhanced an array of techniques that could be exploited to advance food production.

2. Genetic modification technology

The acquisition of new genes that confer selective merits is a relevant factor in genome evolution. Significant proportions of prokaryotic and eukaryotic genomes originated from the exchange of genetic material among related or unrelated species through horizontal gene transfer (HGT). The HGT technique has been noted as one of the key sources of molecular variability and driver of evolution [11, 12]. This HGT often results in the occurrence of crown galls and the mechanism of HGT has been well understood and reported [12–14]. Moreover, Kyndt et al. [12] found that all the 291 tested accessions of cultivated sweetpotato contain one or more transfer DNA (T-DNA) sequences similar to the cultivated sweetpotato clone "Huachano", suggesting that an Agrobacterium infection occurred in evolutionary times. This finding also depicts the importance of plant–microbe interactions, and given that this crop has been eaten for millennia, it might change the negative mindset and paradigm governing the "unnatural" status of transgenic crops. The plant regeneration system of a typical root crop such as cassava was fully developed in the 1990s using somatic embryogenesis, shoot organogenesis from cotyledons of somatic embryos (somatic cotyledons), and friable embryogenic calli (FEC) (**Figure 1**). The key media components used for the establishment and optimization of the plant regeneration system have been well noted by Liu et al. [15]. The common transgenic techniques utilized for the genetic transformation of root and tuber crops are Agrobacterium-mediated gene delivery and particle bombardment. The explants used for transformation include somatic cotyledons and FEC.

2.1 *Agrobacterium*-mediated genetic modification method

Agrobacterium-mediated genetic modification has been the method of choice for the development of genetically modified crops. The major merits of Agrobacterium-mediated genetic modification are its high frequencies of single-copy integration, high reproducibility, transformation efficiency, stable expression of transgenes, utilization of simple equipment, ease of accessibility, ability to transfer low copies of DNA fragments carrying the desirable genes at higher efficiencies with minimal cost and the transfer of very large DNA fragments with low rearrangement [17].

The first attempt to transform cotyledonous embryos of cassava MPer183 with *Agrobacterium tumefaciens* CIAT1182 started in 1993–1994, but the transgenic

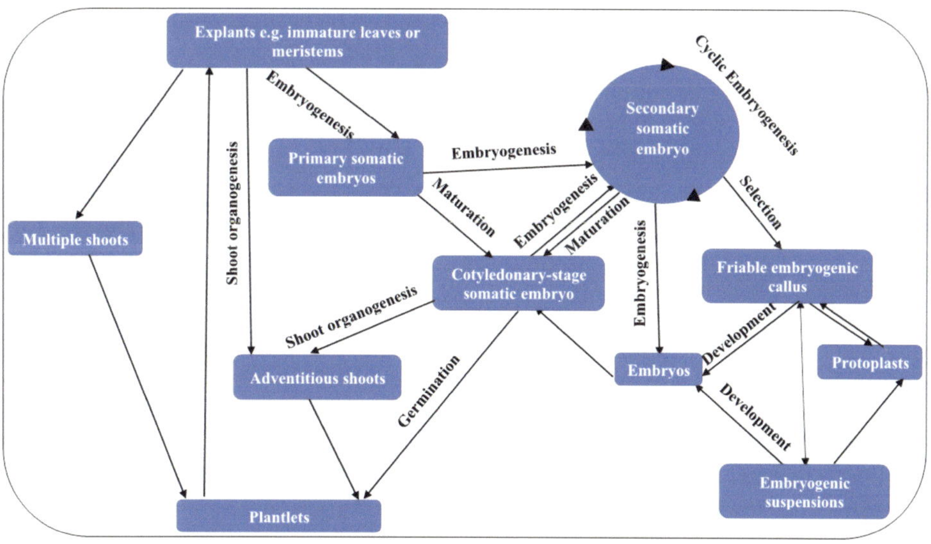

Figure 1.
Illustration of an in vitro plant regeneration system of a typical root crop. Redrawn from Zhang [16].

nature of regenerated plants could not be verified by Southern analysis [18]. The first successful Agrobacterium mediated cassava genetic transformation was done in Potrykus laboratory at ETH Zurich in 1996 [19]. Several Agrobacterium strains sheltering different binary vectors such as LBA4404 (pTOK233), LBA4404 (pBin9GusInt), C58C1 (pIG121Hm), and EHA105 (pBin9GusInt) were used for the transformation of cassava somatic cotyledons. Gonzalez et al. [20] successfully transformed TMS60444 with the *A. tumefaciens* strain ABI using Agrobacterium-mediated FEC transformation. The transgenic nature of the two plant lines resistant to paromomycin was confirmed by glucuronidation glycosidase (GUS) assay and Southern analysis. The embryogenic suspensions of TMS60444 were transformed with *A. tumefaciens* LBA4404 using negative- and positive-selection agents [21]. A total of 12 morphologically-normal transgenic lines have been developed, of which, are five are mannose selected and seven hygromycin resistant. Moreover, polymerase chain reaction (PCR) and Southern analysis confirmed the successful integration of the transgene into the genome. The confirmation of expression of the transgene in the regenerated plants was done using reverse transcription (RT)–PCR and Northern analyses. In 2000, Sarria et al. [22] successfully transformed a herbicide (phosphinotricin, ppt)-resistance gene into the cotyledons of cassava MPer183 by an Agrobacterium-mediated method and found stable transgenic plants resistant to Basta spray (at concentrations of 200 mg/L or more). An efficiently robust and reproducible transformation protocol for cassava embryogenic suspension culture with *A. tumefaciens* has also been reported [23]. Of the 31 GUS-active plants identified, 14 were found with 100% GUS activity, whilst the remaining lines had 72% GUS activity. The transgenic nature of these plants was detected using the southern blot analysis. Zhang et al. [24] successfully introduced a synthetic artificial storage protein 1 (ASP1) gene encoding a storage protein rich in essential amino acids (80%) into embryogenic suspensions of cassava using Agrobacterium-mediated FEC transformation. The ASP1 tetramer was detected in the leaves and primary roots of transgenic cassava plants by Western analysis. Another achievement in the cassava genetic modification is the development of transgenic cassava with a lower cyanide content using MCol2215 cotyledon explants [25]. Jørgensen et al. [26] constructed several RNAi and antisense vectors to interfere with the expression of CYP79D1 and AYP79D2, and transformed the somatic cotyledons of cassava MCol22 using an Agrobacterium-mediated technique. In 2009, a multi-auto-transformation (MAT) vector system of isopentenyl transferase (ipt) type was utilized for the production of marker-free transgenic cassava plants with conversion efficiency up to 19–21% via shoot organogenesis of KU50 somatic cotyledons [27]. Zhang et al. [28] developed transgenic cassava with senescence-inducible expression of the ipt using Agrobacterium-inoculated TMS60444 somatic cotyledons via shoot organogenesis. These achievements contributed to cassava transformation for the verification of a tissue-specific promoter [29, 30], the resistance to African cassava mosaic virus (ACMV) [31], increased protein content [32], and improved cassava brown streak virus resistance [33].

For sweetpotato, Kyndt et al. [12] reported that the Agrobacterium-mediated gene delivery system was utilized for T-DNA integration, the interruption of an *F-box* gene, and the subsequent insertion of foreign T-DNA into the sweetpotato genome. This is believed to have occurred during the evolution and domestication of this crop. White et al. [34] suggested that the identification of gene sequences in IbT-DNA1 and IbT-DNA2 imply that the transforming Agrobacterium was probably Agrobacterium rhizogenes, an ancestral form of A. rhizogenes or a closely related species (perhaps extinct) of A. rhizogenes. The *Ib*T-DNA1 corresponds to TR-DNA (typically containing the auxin biosynthesis genes *iaa*M and *iaa*H), and *Ib*T-DNA2 corresponds to TL-DNA (harboring the *Rol* genes). The gene organization and DNA sequences of the T-DNAs are similar to, but distinctly different from, the ORFs of the Ri and Ti-plasmids in well characterized *Agrobacterium* strains. The identified *RolB/RolC* region represents a new member of the *RolB* family indicating that, unlike the T-DNA found in *Nicotiana* spp. [35], the *Agrobacterium* strain (or species) that transferred its T-DNA into the sweetpotato genome is uncommon.

For yams, the Agrobacterium-mediated gene delivery system is the most preferred technique utilized for the genetic modification of the crop [14, 36]. Initial development of a transient genetic modification of *Dioscorea rotundata* using the Agrobacterium-mediated produced no transgenic plants [37]. However, the first fast, efficient and reproducible protocol for Agrobacterium-mediated transformation of *D. rotundata* resulted in the generation of stable transformations and the regeneration of complete transgenic yam plants [8]. This achievement laid the foundation for the full implementation of genetic engineering and gene editing in yam. Based on the review, it is clear that Agrobacterium-mediated transformation system is the most prominent genetic modification technique due to the availability of a large number of transgenic plants. The transformation efficiency of this technique can be improved using a protocol based on somatic cotyledons as explants for the transformation of cassava, sweetpotato and yams. Thus, it is the most widely-used method for genetic engineering in the studied root and tuber crops.

2.2 Biolistic-mediated genetic modification method

Biolistic transformation or gene gun or particle bombardment technique is often utilized for plant transformation studies. The technique involves FEC induction, subculturing, somatic embryogenesis, and plant germination. The biolistic transformation technique involves series of protocols and stages [15]. These include the subculturing in SH liquid medium, followed by supplementation with 50 μM picloram, without selection for 2 weeks. The samples are further subjected to SH liquid medium with 25 μM paromomycin for 4–5 weeks, solid SH medium with 25 μM paromomycin for 4 weeks, and Murashige and Skoog (MS) medium with 5 μM picloram for embryogenesis. The next stage involves the development of transformed cell clusters into somatic embryoids in maturation media (MS medium supplemented with 0.5% activated charcoal). The final stage is the regeneration and verification of transgenic plants exhibiting GUS-positive and paromomycin resistance using Southern analysis [15]. This technique was successfully conducted on somatic cotyledons of cassava genotypes CMC40, MPer183, MCol22, and TMS60444 suspensions [21, 38]. Twenty transgenic plants of TMS60444 and 11 transgenic lines of MCol22 were obtained. The GUS, Southern blot, and RT-PCR assays indicated the successful integration of the transgene into the plant genome. Transgenic cassava has also been produced from the plasmid constructs pHB1 and pJIT100 using FEC of TMS60444 and particle bombardment [39]. Of the dozens of transgenic plants produced using the pHB1 and pJIT100 constructs, some have been analyzed at the molecular level. Zhang and Puonti-Kaerlas [40] used particle bombardment to transfer the plasmid pHMG into TMS60444 embryogenic suspensions. Selection from the dozens of transgenic cassava plants produced in less than 15 weeks was based on either negative hygromycin or positive mannose. Zhang et al. [41] also utilized FEC and a particle bombardment technique to investigate ACMV resistance in cassava.

The success of this method depends on the high efficiency of the particle bombardment and shoot organogenesis. The long time utilized for the FEC induction, subculturing, somatic embryogenesis, and plant germination causes a low efficiency of plant regeneration and a high rate of somaclonal variation. This indicates low probability of success due to its complicated operation and its susceptibility to many factors. Thus, the technique is infrequently used for genetic transformation in roots and tubers. This necessitates the establishment of a robust standard protocol for FEC-based transformation for each cultivar.

2.3 The friable embryogenic callus genetic modification method

The embryogenic callus tissue utilized in this genetic modification technique are friable as they tear, slough and bleed more easily when touched. For the cassava embryogenic callus genetic modification, the induction of primary somatic embryos on the embryogenesis induction medium of the immature young leaves and apical or axillary meristems are useful for the establishment of cyclic secondary somatic embryogenesis using the subculture of the induction medium [15]. However, continual subculturing of the secondary somatic embryos on Gresshoff and Doy (GD) medium, supplementation with 12 mg/L picloram results into production of FEC, formation of non-embryogenic calli and secondary somatic embryos [15]. Establishment of an embryogenic suspension for rapid multiplication in liquid Schenk and Hildebrandt (SH) medium containing 10–12 mg/L picloram requires appropriate selection of FEC and subculturing on the GD solid medium in the long-term. Culturing of embryogenic suspension cells on MSN solid media with 1 mg/L naphthaleneacetic acid, produce somatic embryos and subsequently cotyledon-stage somatic embryos that germinate to plantlets [42, 43]. The FEC and embryogenic suspension cultures are susceptible to Agrobacterium infection, and favorable to particle bombardment, making the delivery of foreign genes easy. Cassava transformation is usually done using FEC and/or embryogenic suspensions by *A. tumefaciens* or particle bombardment.

2.4 Shoot organogenesis-based genetic modification method

This technique involves the induction of somatic embryogenesis from immature leaves and apical or axillary meristems using green cotyledons as explants and plant regeneration via shoot organogenesis to overcome genotype restrictions [44]. However, an efficient shoot organogenesis using mature green cotyledons has been developed from secondary somatic embryos in AgCOM medium supplemented with N6-benzylaminopurine and $AgNO_3$ (an ethylene action inhibitor) [45]. The shoot organogenesis-based genetic modification method is most applicable and suitable for the biolistic or Agrobacterium-mediated genetic transformation since these techniques accommodate the initiation of shoot primordia from cut ends and epidermal cells as well as the regeneration of transgenic plants in a short cycle of 3–4 months. Zhang et al. [21] established that subjection of transgenic shoots to rooting sensitivity tests could quickly eliminates false-positive transgenic plants, thereby preventing difficult and expensive molecular analyses. This is also a reliable method for screening transgenic cassava plants.

3. Novel genetic transformation technologies

3.1 Genetic transformation

Advances in transgenic technology has led to the development of different schemes for the delivery of target genes into plant cells [15]. Selectable-marker genes including herbicide- and antibiotic-resistance genes, and the reporter genes such as GUS, luc, chloramphenicol acetyltransferase, and green fluorescent protein. The selectable-marker and reporter genes are used for screening and monitoring of putative transgenic plants. However, there are lots of public concern about the biosafety of using these marker genes are widely used for genetic transformation. This necessitates use of suitable alternatives as selectable markers. Roots and tubers are conventionally vegetatively propagated crops with the merit of reduced risk of horizontal gene transfer to relevant organisms and pathogens through pollen. This indicates the necessity of developing safe marker genes or marker-free technology in root and tuber crops.

3.2 Marker-free technology

Development of marker-free transgenic plants involves co-transformation, homologous recombination, site-specific recombination, and transposition systems [15]. These techniques have potential for application in cassava, sweetpotato and yams once the genetic transformation systems of the crops are further improved and optimized.

3.3 Co-transformation system

The co-transformation system involves the co-transformation of plant cells with a pair of plasmid vectors, each carrying a selectable-marker gene or a target gene. The selectable marker and target gene integrate into different loci on the chromosome at the same period. Since the two genes integrate at different loci, the selectable-marker gene segregates to produce marker-free transgenic plants. This procedure works for sexually producing crops [46]. For asexually reproducing crops, conventional breeding is utilized for development of new varieties. Thus, co-transformation via sexual hybridization technique is a useful means of producing marker-free cassava, sweetpotato and yam transgenic plants as an intermediate putative parental genotype. The major demerits of this method include its labor intensiveness and the long cycle for the occurrence of genetic transformation [15].

3.4 The recombination system

The site-specific recombination systems reported include Cre/loxP (Cre: causes recombination; loxP: locus of crossing X2 over in P1), R/RS (R: recombinase; RS: recognition site), Gin/gix (Gin: inversion of the G loop; gix: Gin-inversion complex sites), and FLP/FRT (FLP: flipping DNA; FRT: FLP recombination target), consist of a recombinase enzyme and corresponding specific recognition sequences [15]. The functions of the recombinase enzyme are to identify and mediate the recombination of two specific recognition sequences in the same direction leading to the simultaneous and independent formation of the cyclic DNA and the chromosome [15]. This facilitates the elimination of the selective marker gene. Of the three recombination systems, Cre/loxP is the most widely used and studied system. Dale and Ow [47] first utilized the Cre/loxP system to develop transgenic tobacco. The application of site-specific recombination system is still in its infancy stage in root and tuber crops due to its lower transformation efficiency. Saelim et al. [27] reported marker-free transgenic KU50 cassava produced using the MAT vector system (containing the yeast site-specific recombination Rint/RS system mediated excision of DNA fragments and the ipt phenotypic marker gene from recombination sites) developed by Ebinuma et al. [48]. The development of excessive and overgrowth buds in transgenic plants is caused by isopentenyl transferase gene. Genetic recombination during subculture eliminates the expression of this gene to produce marker-free transgenic plants. The conversion efficiency and the proportion of normal growing plants are 19–21% and 32–38%, respectively, indicating feasibility of its use. Zuo et al. [49] developed the Cre-loxP-XVE system and could be worth testing in root and tuber transgenic program.

3.5 The transposon system

Transposable elements (TEs) or jumping genes or transposons are sequences of DNA that move from one site in the genome to another. The transposons take advantage of their characteristic conservative cut-paste mechanism to transfer genes from one site to another in the chromosome [15]. This attribute of transposons reduces the probability of genetic linkage at the new site even after occurrence of transposition to the new site. This peculiar attribute of TEs is useful for the removal of unwanted marker gene from the transgenic plants. The removal of the unwanted marker gene is accomplished in two ways: (i) the marker gene could be placed between the TEs and the repeat sequence Dissociator (Ds) element; or (ii) the target gene could be placed between the Ds sequences. During transposition, the marker gene maybe lost or separates from the target gene. The occurrence of the transposon is achieved by progeny segregation; therefore, its application in root and tuber crops is very difficult.

3.6 Non-antibiotic selection

In the non-antibiotic selection system, growth of transformed cells is favored by the development, promotion or additional metabolic activity, without affecting non-transformed cell growth by hormonal stimulation or inhibition by starvation from nonmetabolizable sugars, or death by antibiotics [15]. Unlike the conventional antibiotic-selection systems, the non-antibiotic techniques exhibit better acceptability by the public as they are considered to be safer than the antibiotic-selection systems. Selectable-marker genes from hormonal action Isopentenyl transferase (ipt) and indole-3-acetamide hydrolase (iaaH) are reportedly the most widely used hormone metabolism genes [15]. The Isopentenyl transferase gene was cloned from *A. tumefaciens* T-DNA and is related to cytokinin biosynthesis. The overexpression of ipt in transgenic plants has been observed to cause phenotypic variations in them [15]. This necessitates use of an inducible promoter to regulate its expression or the ipt gene should be applied in a site-specific recombination enzyme system or transposon system to build a highly-efficient marker-removal system. Saelim et al. [27] opined the usefulness of the Rint/RS system for excision of the ipt gene from transgenic cassava KU50. This technology is useful in the genetic transformation of crops with long cycle of breeding. The iaaH is also a safe selective-marker gene detected in the regulation of hormone metabolism. The overexpression of iaaH results in abnormal production of transgenic plants. However, the expressivity of iaaH can be modulated, inactivated or removed [15].

3.7 Selectable-marker genes based on sugar metabolism

The selectable genes consist of selectable markers utilized to facilitate the isolation of plasmid-containing transformants [15]. Selectable markers are useful for genetic transformation as they permit plant cells to grow under conditions that prevent the growth of untransformed tissue. Selectable-marker genes based on sugar metabolism comprise of the mannose phosphate isomerase (pmi) gene and the xylose isomerase (xylA) gene [50]. The pmi system is a positive selection system that uses D-mannose as a selection agent. The mechanism of the system is well reviewed by Liu et al. [15]. The pmi gene has been widely applied in the transformation systems of rice, corn, wheat, and sugar beet [51]. The pmi-mannose system has also been used in cassava transformation system [40]. For validation of this technique, Zhang [16] constructed a pHMG binary vector that uses the visual marker GUS gene, independent expression cassettes of the pmi and hygromycin phosphotransferase genes in its T-DNA region. They observed 82.6% Agrobacterium-mediated transformation of embryogenic suspensions by the mannose selection system compared to 100% hygromycin selection system. Transgenic plants were also generated from application of the biolistic transformation of somatic cotyledons as explants, and mannose as the selective agent [40].

Another selectable marker xylA gene, encodes xylA and catalyzes the conversion of D-xylose through D-xylulose tautomerism. The growth of transformed cells is promoted in the medium containing a carbon source material such as D-xylose, whereas the growth of non-transformed cells is inhibited by the lack of a suitable carbon source [15]. Haldrup et al. [52] noted the production of transgenic plants using D-xylose as the carbon source. Application of this technique is yet to be fully exploited in root and tuber crops.

3.8 Tissue-specific promoter

Liu et al. [15] reported three types of promoters based on spatial and temporal expression patterns including inducible promoters, constitutive promoters, and tissue- or organ-specific promoters. For cassava genetic transformation, the CaMV 35S promoter and methyl jasmonate and salicylic acid-induced nopaline synthase gene promoter are the widely used promoters. These promoters regulate genes of interest and selective marker gene or reporter gene. The gene expression level of CaMV 35S promoter in transgenic cassava leaves was noted to be higher relative to the storage roots indicating possible influence on the function of certain genes in cassava storage roots [29].

Further studies to increase the specific expression of a target gene in cassava organs and tissues led to the discovery of specific promoters in leaves or storage roots. Zhang et al. [29] detected two cassava promoters such as p15/1.5 of a cytochrome P450 protein, and p54/1.0 of the cassava glutamic acid-rich protein, Pt2L4, from a cassava storage root cDNA library. The gene expression patterns of these promoters show close association with cassava vascular tissues and storage root development, especially in the starch-rich parenchyma cells of the storage root. Their activities are also stronger than those of CaMV 35S promoter. The function of the glutamic acid-rich protein promoter has been well articulated [30, 53–55]. Application of p54/1.0 promoter regulated the dsRNA expression for interference with the granule-bound starch synthase (GBSS)I expression producing amylose-free transgenic cassava [56].

Tuber-specific class I patatin promoter was noted to regulate several gene expression. Ihemere et al. [57] reported its gene regulation activity in the expression of the *Escherichia coli* gene, glucose-1-phosphate adenylyltransferase (glgC) with insensitivity to the substrate, by site-directed mutation in cassava for increased starch content. Siritunga and Sayre [25] noted its regulation activity in CYP79D1 and CYP79D2 antisense gene expression in cassava for decreased toxicity of cyanide. Abhary et al. [32] reported the role of the promoter in the expression of a zeolin fusion protein to increase protein content. A promoter regulating protein family AAI_LTSS of unknown function in cassava was found strongly expressed in the secondary xylem of the carrot [58].

Leaf senescence inducible promoter SAG12 was found to regulate the ipt gene that moderate extended leaf longevity and improved the drought resistance in transgenic [28]. Leaf specific cab1 promoter applied in transgenic cassava regulated transgene expression Siritunga and Sayre [25]. Based on existing information on tissue-specific promoters, greater application of this technology is envisaged in future root and tuber breeding programs.

4. Applications of genetic modification technology in cassava, sweetpotato and yam improvement programs

Application of functional genomics and genetic engineering has contributed to resolve the problems associated with the germplasm enhancement of root and tuber crops. For cassava, improved agronomic traits achieved by transgenic technology include virus resistance, improved nutritional quality, reduced cyanide content, improved biomass, and delayed post-harvest physiology deterioration in storage roots.

4.1 Resistance to pests and viral diseases

Pests and viral diseases are transmitted via infected stems, vines, tubers from generation to generation, subsequently causing yield losses. For instance, the cassava mosaic disease (CMD) accounts for about 20–95% yield reductions [59]. It is the major cassava disease in Africa and the Indian peninsula. The CMD is caused by several cassava geminiviruses and their satellite components, including the ACMV, Eastern ACMV, and Indian cassava mosaic virus [59, 60].

Transgenic technology has played an important role in obtaining virus resistant cultivars [61]. Chellappan et al. [62] utilized pILTAB9001 and pILTAB9002 harboring the wild-type and mutant AC1 genes of ACMV-Kenya. These wild-type and mutant AC1 genes regulate the cassava vein mosaic virus promoter and the pea Rubisco terminator, which enables the production of transgenic TMS60444 plants with increased resistance to mosaic disease [62]. The initial inoculation assay detected transgenic plants resistant to several cassava geminivirus diseases from Africa. However, these transgenic plants were later susceptible to CMD infection in a closed-field trial in Kenya. With the aid of improved antisense RNA technology, Zhang et al. [63] developed transgenic cassava plants with increased ACMV resistance targeting the viral mRNAs of Rep (AC1), TrAP (AC2), and REn (AC3). Several transgenic clones remained symptomless after biolistic inoculation of ACMV at infection pressure of 100 ng viral DNA plant^{-1}. Decreased and attenuated symptom development were also detected even at higher viral DNA doses. Significant reduction in viral DNA accumulation was observed in the leaves of transgenic ACMV-resistant plants. Application of RNAi-mediated gene-silencing approaches, siRNAs, homologous to either the common region or AC1 in transgenic cassava plants suppressed the replication of African mosaic virus, leading to recovery after infection with ACMV [64] or immunity to infection by the virus [31]. Bi et al. [65] screened cassava germplasms from using Agrobacterium-mediated inoculation in combination with CMD-resistant molecular markers RME1, SSRY28, and NS158. Findings showed that the cassava germplasms lacked CMD-resistant genes, suggesting the necessity of introducing and integrating disease-resistant cassava genotypes from Africa into the current breeding program, while developing CMD-resistant cassava using different transgenic approaches.

Cassava brown streak disease (CBSD) is another important viral disease of cassava in Africa. A sequence analysis of CBSD showed that the causative virus belongs to potato virus-Y of the ipomovirus family [66]. Two subspecies of CBSD are cassava brown streak virus (CBSV) and cassava brown streak Mozambique virus (CBSMV) [67]. Transgenic approach is noted to be more promising for development of CBSD clones relative to the traditional cassava-breeding method. Resistant cassava lines have been developed by transferring the virus coat protein gene or through RNA (siRNA) interference [33, 68]. Cassava bacterial blight (CBB) disease (gum disease), caused by *Xanthomonas manihotis*, mainly affects cassava leaves. Transcriptomic studies have shown a rapid change in cassava genes after infection by this disease [69]. The main insects that infest cassava are whiteflies, cassava mealybugs, cassava green mites, and stemborers, while root-knot nematodes are the most widely-reported parasitic nematodes on cassava. The cultivation of insect resistant cassava genotypes increases the yield and quality of the crop. Insecticide proteins including Bt Cry proteins, protease inhibitors, α-amylase inhibitor, and plant lectins, could aid insecticides, as a high expression of these products in transgenic cassava might facilitate increased insect resistance.

Targeted genome alteration technique is a promising tool for yam breeding. Successful application of the CRISPR/Cas9 technology resulted into inactivation of the endogenous banana streak virus by editing the virus sequences to develop resistant plantain [70]. Yam viruses have also been found to be integrated into the genome of yam and for the development of yam genotypes resistant to yam mosaic virus using CRISPR/Cas9 approach [71, 72]. The major challenge of the CRISPR/Cas9 technology is that it may recognize sequences with up to five mismatched bases suggesting high rates of off-target effects [73]. However, techniques such as DNA-RNA chimeric guides, Cpf1, a single RNA endonuclease that employs a T-rich PAM on the 5′ side of the guide, and specific point mutations have been developed to mitigate this challenge [74, 75].

The recently established gene-editing technique, the clustered regularly interspaced short palindromic repeats (CRISPR/Cas9) system, resulting from the adaptive immune system of *Streptococcus pyogenes*, is a notably potent tool for targeted genome editing in many species [76]. Gene editing and genetic engineering technologies have been reported to facilitate improvement of the productivity and nutritional quality of yam. This was achieved through the Genome-Enabled Platforms for Yam Project launched in 2016 in collaboration between scientists at the International Institute of Tropical Agriculture (IITA) and Iowa State University (https://www.nsf.gov/awardsearch/showAward?AWD_ID=1543888). Moreover, a genome-editing tool for yam using phytoene desaturase (a key enzyme in the β-carotene biosynthesis pathway, which converts the colorless phytoene to colored carotenoids) as a marker is being developed [77]. Targeted traits such as resistance to yam mosaic virus and anthracnose diseases, herbicide tolerance and nematode resistance are being investigated using yam gene editing and genetic engineering technique.

Feng et al. [76] successfully applied the CRISPR/Cas9-mediated targeted mutagenesis in *D. zingiberensis* using an *A. tumefaciens*-mediated transformation method. Their study aimed at an essential gene involved in the synthesis of secondary metabolites, known as the farnesyl pyrophosphate synthase gene (Dzfps). They detected five types of mutations among the transformed plants at the predicted double-stranded break site. Feng et al. [76] also found that the transcript levels of Dzfps and the content of squalene in isolated mutants were drastically decreased relative to those in wild-type plants and concluded that CRISPR/Cas9 is a rapid and efficient method for targeted genome modification in *D. zingiberensis*.

4.2 Improved stress resistance

Cassava is susceptible to cold stress, resulting into decrease root yield and cumbersome maintenance of cassava stems. In southern China, winter, rain, snow, and freezing temperatures cause severe frostbite to cassava stems [15]. In 2008, damage of cassava stems caused by cold stress in Guangxi amounts to several billion

RMB. Thus, development of low temperature resistant cassava clones aids easier stem storage and increasing cultivation in temperate countries, thereby meeting the demand for raw cassava materials for the industries. High expression of the C-repeat/dehydration-responsive element-binding factor 3 gene regulated by a low temperature inducible or CaMV 35S promoter has been noted to significantly improve cold resistance in transgenic cassava [15]. The growth and yield of cassava are influenced by intermittent drought in tropical and subtropical regions. Application of leaf senescence induced promoter, SAG12, for expression of the ipt gene, transgenic cassava revealed prolonged leaf life, and improved resistance to drought stress [28]. This study demonstrates a useful strategy for the improvement of drought resistant and high yield cassava clones. Several authors have also noted a large number of genes associated with stress resistance using high-throughput genomic and transcriptomic analyses of cassava [78–80]. Functional verification of these genes is needed for better understanding of the molecular mechanisms of cassava resistance to different stresses and establishment of the theoretical basis for cassava genetic improvements.

4.3 Enhanced nutritional quality

Nutritional quality enhancement in roots and tubers can improve the dietary and nutritional balance of people who utilize them, particularly in central and western Africa. Zhang et al. [24] noted the production of transgenic cassava exhibiting an artificial storage protein enriched with essential amino acids, under the control of the CaMV 35S promoter. The transgenic plants were detected expressing ASP1 at both the RNA and protein levels. The leaves of transgenic plants had increased proline and serine contents, while the amounts of aspartic acid, alanine, and methionine were decreased compared to non-transgenic plants. Altered composition of amino acids and reduced cyanide content were also observed in transgenic plants [32]. Thus, cassava protein enhancement using transgenic approach is practically feasible and is a useful means of reducing protein deficiency in poverty-stricken regions [81].

The BioCassava Plus project team has developed transgenic cassava clones with value-added traits including virus resistance, improved protein content, and increased vitamin A, iron, and zinc contents [10]. The HarvestPlus project team has also developed β-carotene-rich cassava clones [82].

4.4 Reduced cyanide content

Siritunga and Sayra [25] utilized an Arabidopsis leaf-specific promoter to drive the antisense expression of cytochrome P450 genes (CYP79D1 and CYP79D2). In vitro tests of the linamarin content of the transgenic leaves revealed a decrease of 60–94% compared to the control, while a 99% decrease was observed in the storage roots of cassava, suggesting the transport of linamarins from leaves to storage roots. White et al. [83] noted that at transcript level, the hydroxynitrile lyase content in cassava roots is only 6% of that in the leaves. The overexpression of hydroxynitrile lyase reduces the acetone cyanohydrin content of roots, thereby accelerating the detoxification process. The overexpression of hydroxynitrile lyase was achieved through the cloning of the cDNA of the gene encoding hydroxynitrile lyase between the CaMV 35S promoter and the pea ribulose bisphosphate carboxylase terminal sequence, and transformed into MCol2215 [84]. The authors found a 40–135% increase in the hydroxynitrile lyase activity in transgenic plants, compared to 800–1300% found in the storage roots of cassava. However, no changes were detected in the total amount of linamarin and lotaustralin detected in the whole plant [84]. After harvesting, the detoxification capacity of the root was greatly enhanced. Jørgensen et al. [26] conducted similar trial by RNAi and found a 92% decrease in the cyanogenic glucoside contents of cassava storage roots.

4.5 Improved starch content and quality

Starch quality is one of the key agronomic traits for selection of elite cassava genotypes. The ratio of amylose to amylopectin determines the property of starch granules, and influences the quality of various starch products utilized in the pharmaceutical, chemical, and paper-making industries. Starch synthesis is regulated by AGPase, starch synthase (SS), and starch-branching enzyme (SBE). These three main enzymes have been successfully cloned from cassava [85–87]. The inhibition of AGPase activity results in partial or complete termination of starch synthesis. Thus, improvement of the AGPase activity contributes to the conversion of sugar to starch, which subsequently increases the starch quantity. The reduction of allosteric feedback regulation by fructose-1,6-bisphosphate was detected by genetic modification of the *E. coli* glgC gene (encoding AGPase) by site directed mutagenesis (G336D) [57]. Transgenic cassava plants expressing the mutant glgC gene showed a 70% enhancement in AGPase activity, and up to a 2.6-fold increase in biomass. The quality of cassava starch depends on amylose and amylopectin content. Antisense RNA technology has been used to reduce GBSS expression in the potato resulting into decreased amylose content in potato tubers [88]. Using antisense GBSSI RNA under the control of the CaMV 35S promoter, waxy transgenic cassava plants were generated [89]. Waxy transgenic cassava clones have also been developed using the cassava vascular-specific p54/1.0 and CaMV 35S promoters to drive the expression of hairpin dsRNA homologous to cassava GBSSI. Starches from waxy transgenic plants revealed altered biological and physico-chemical properties [56]. Thus, the control of GBSS activity is an effective way to regulate amylose synthesis.

4.6 Delayed post-harvest physiological deterioration

Postharvest physiological deterioration (PPD) limits the storability and utilization of cassava. The PPD is a physiological and biochemical decay process caused by an oxidative burst in storage root cells of cassava [90]. The PPD phenomenon has a close relationship with reactive oxygen species (ROS) [90]. Analysis of the proteins and enzymes influencing PPD using the cDNA-AFLP technique showed that most of the proteins and enzymes are involved in signal transduction, ROS, cell wall repair, programmed cell death, metabolite transport, signal transduction, and a series of biological processes [91]. The upregulation or downregulation of key enzymes or factors in the PPD pathway by the overexpression or RNAi might effectively slow or decrease the PPD activity [91]. Study of the temporal and spatial expressions of genes related to ROS production and scavenging in cassava PPD, and the functional verification of key genes, indicates the possibility of interference of the PPD process by the regulation of ROS-scavenging activities [15]. These genes are involved in the regulation of glutathione-peroxidase cycle, the ascorbate-glutathione cycle, and the peroxidase-oxidoreductase cycle [15]. The genes also combine with superoxide dismutase to generate univalent, bivalent, and trivalent overexpression or RNAi vectors to transform cassava. Secondary metabolites of PPD such as scopoletin and diterpenoids, are synthesized in deteriorated roots [92]. The interference with the biosynthesis of the secondary metabolites might also influence PPD. Morante et al. [93] noted the discovery of germplasms of radiation mutants that suppress the occurrence of PPD. These results provide gene resources for the amelioration of PPD through traditional and molecular breeding techniques.

5. Future prospects of genetic modification for root and tuber improvement

Pests and diseases are among key factors affecting yield and quality of root and tuber crops. The progress made in genetic modification of these crops could open up many avenues to produce disease resistant varieties, through pathogen-derived resistance strategies, that would not be possible using conventional breeding approaches alone. In yams, host plant resistance to anthracnose has been suggested as a more viable alternative to control yam anthracnose disease (YAD) than use of chemical fungicides [94]. However, studies reveal lack of genotypes resistant to the disease [94]. Thus, the most efficient strategy for YAD control is possibly the development of disease resistant plants using the transgenic approach. These approaches could include the expression of genes encoding elicitors of defense response [95], genes encoding plant, fungal or bacterial hydrolytic enzymes [96] and antimicrobial peptides (AMPs) [97]. Most AMPs are non-toxic to plant and mammalian cells, with a broad-spectrum antimicrobial activity against fungi and bacteria.

Use of nematode resistant yam varieties can be an effective strategy in controlling the disease, however, there are no resistant varieties to nematodes. The application of transgenic approach could serve as a viable alternative for improvement of the nematode resistance of yam. Several transgenes have been noted to confer plant resistance to both tropical and temperate plant parasitic nematodes [98]. Cystatins inhibit nematode digestive cysteine proteinase activity, thereby suppressing the growth and multiplication of these pests [99]. Cystatin is one of the transgenes that has been successfully applied to control plant nematodes. The cystatins transgene confers improved resistance to a range of nematodes in different crops including potato, sweetpotato, rice, tomato, and plantain [100–105]. The transgene has displayed proven efficacy under field conditions [102]. Such an approach could be exploited for genetic enhancement of resistance of yam against nematodes in the near future.

In sweetpotato, genetic modification studies show that the suppression of βcarotene hydroxylase (CHY-β), which catalyzes the hydroxylation steps of both β-carotene into β-cryptoxanthin and β-cryptoxanthin into zeaxanthin, significantly increased the β-carotene and total carotenoid content in transgenic cultured cells [106]. Moreover, suppression of lycopene β-cyclase (LCY-β), which catalyzes the cyclization steps of lycopene to produce β-carotene, resulted in increased total carotenoid content [107]. Transgenic sweetpotato plants overexpressing an Or homolog, IbOr, showed increased carotenoid contents compared to non-transformed control plants [108]. These findings indicate that the CHY-β and LCY-β are key enzymes of carotenoid biosynthesis in sweetpotato that could be exploited as good targets for molecular breeding.

In many plants, transcription factors belonging to MYB-family are implicated in the control of a tissue-specific accumulation of anthocyanin. Mano et al. [109] investigated gene expression analysis using purple-fleshed clones and transformation using sweetpotato leaves and calli. Found that one of the MYB-type transcription factors in sweetpotato, IbMYB1, regulates anthocyanin accumulation in storage roots. The overexpression of this gene successfully induces anthocyanin accumulation in the storage roots of an orange-fleshed sweetpotato cultivar, resulting in higher radical scavenging activity [110]. Comparison of the structure of the IbMYB1 genes between high-anthocyanin content and non-anthocyanin cultivars revealed distinct IbMYB1 copies of IbMYB1-2a and IbMYB1-2b, shared only in the high-anthocyanin cultivars and their common ancestor 'Yamagawamurasaki' [111]. The PCR fragment amplified from IbMYB1-2a and IbMYB1-2b cosegregated with the pigmentation of the storage roots in the F1 progenies of high-anthocyanin cultivars, suggesting the usefulness of the PCR fragment as a selection marker for high-anthocyanin lines [111].

Transgenic sweetpotato plants overexpressing IbMYB1 were detected to exhibit an elevated total polyphenol level [110]. The gene expression of phenylalanine ammonia lyase (PAL), cinnamate 4-hydroxylase (C4H), and 4-hydroxycinnamoyl-CoA ligase (4CL), involved in the early steps of both anthocyanin and caffeoylquinic acids (CQA) biosynthesis of the transgenic plants, was upregulated. Tanaka et al. [111] also noted a suppressed expression of these genes in a white-fleshed mutant of 'Ayamurasaki'. Padda and Picha [112] detected variations in the CQA content in the storage roots of non-purple fleshed cultivars, leading to the speculation that, in addition to a co-regulation of the anthocyanin content, a specific regulation mechanism of CQA content exists.

6. Conclusion

Successful genetic modification and application in complementarity with conventional population improvement techniques and advanced genomics and phenomics tools can contribute and accelerate the genetic gain in cassava, sweetpotato and yams compared to using conventional technique only. Genetic modification technology in root and tuber crops is transitioning from development to application. This has great potential in promoting industrialization of these crops and their immense contribution to food security. Some of the successes of transgenic technology include virus resistance, improved nutritional quality, improved starch yield and quality, reduced cyanide content in cassava, improved biomass, and delayed post-harvest physiology deterioration in storage roots. Based on existing literature, genome editing should be incorporated into the root and tuber improvement programmes and targeted traits should be decided in consultation with breeders. The ethics and regulation of genetically modified and gene-edited crops should be seriously considered in the application of these technologies.

Acknowledgements

The authors are thankfully grateful to scientists at the International Institute of Tropical Agriculture (IITA) and lecturers at the University of Ghana whose supports have been tremendous in backstopping our breeding activities.

Conflict of interest

The authors declare no conflict of interest.

Author details

Prince Emmanuel Norman[1*], Daniel K. Dzidzienyo[2], Kumba Yannah Karim[3] and Aloysius A. Beah[4]

1 Sierra Leone Agricultural Research Institute (SLARI), Freetown, Sierra Leone

2 West Africa Centre for Crop Improvement (WACCI), College of Basic and Applied Sciences, University of Ghana, Accra, Ghana

3 School of Agriculture, Njala University, Njala, Sierra Leone

4 School of Natural Resource Management, Njala University, Njala, Sierra Leone

*Address all correspondence to: pnorman@wacci.ug.edu.gh; p.norman@slari.gov.sl; penorman2008@yahoo.com

References

[1] Akanbi WB, Adeboye CO, Togun AO, Ogunride JO, Adeyeye SA. Growth, herbage and seed yield and quality of *Telfairia occidentalis* as influenced by cassava peel compost and mineral fertilizer. World Journal of Agriculture Science. 2007;**3**:508-516

[2] Iyagba AG. A review on root and tuber crop production and their weed management among small scale farmers in Nigeria. ARPN Journal of Agriculture and Biological Science. 2010;**5**:52-58

[3] Norman PE, Asfaw A, Tongoona PB, Danquah A, Danquah EY, Koeyer DD, et al. Can parentage analysis facilitate breeding activities in root and tuber crops? Agriculture Journal. 2018;**8**(95): 1-24. DOI: 10.3390/agriculture8070095

[4] Berloo R, Hutten RCB, Eck HJ, Visser RGF. An online potato pedigree database resource. Potato Research. 2007;**50**:45-57

[5] Lebot V. Tropical Root and Tuber Crops: Cassava, Sweet Potato, Yams and Aroids. Wallingford, XIX: CABI; 2009. p. 413

[6] Ceballos H, Okogbenin E, Pérez JC, López-Valle LAB, Debouck D. Cassava. In: Root and Tuber Crops, Handbook of Plant Breeding. New York: Springer; 2010

[7] NAS National Academy of Sciences. Safety of Genetically Engineered Foods: Approaches to Assessing Unintended Health Effects. Committee on Identifying and Assessing Unintended Effects of Genetically Engineered Foods on Human Health. Washington, DC, USA: National Academies Pres; 2004; p. 256. http://www.nap.edu/catalog/10977.html

[8] Nyaboga E, Tripathi JN, Manoharan R, Tripathi L. Agrobacterium-mediated genetic transformation of yam (*Dioscorea rotundata*): An important tool for functional study of genes and crop improvement. Frontiers in Plant Science. 2014;**5**:1-14. DOI: 10.3389/fpls.2014.00463

[9] CGIAR Science Council. Report of the First External Review of the HarvestPlus Challenge Program. Rome, Italy: Science Council Secretariat; 2008

[10] Sayre R, Beeching J, Cahoon E, Egesi C, Fauquet C, Fellman J, et al. The BioCassava Plus Program: Biofortification of cassava for sub-Saharan Africa. Annual Review Plant Biology. 2011;**62**:251-272

[11] Bock R. The give-and-take of DNA: Horizontal gene transfer in plants. Trends Plant Science. 2010;**15**(1):11-22

[12] Kyndt T, Quispe D, Zhaic H, Jarret R, Ghislain M, Liu O, et al. The genome of cultivated sweet potato contains Agrobacterium T-DNAs with expressed genes: An example of a naturally transgenic food crop. PNAS Early Edition. 2015;**112**:5844-5849. DOI: 10.1073/pnas.1419685112

[13] Zupan J, Muth TR, Draper O, Zambryski P. The transfer of DNA from *Agrobacterium tumefaciens* into plants: A feast of fundamental insights. Plant Journal. 2000;**23**(1):11-28

[14] Gelvin SB. Agrobacterium-mediated plant transformation: The biology behind the "gene-jockeying" tool. Microbiology and Molecular Biology Review. 2003;**67**(1):16-37. DOI: 10.1128/MMBR.67.1.16-37.2003

[15] Liu J, Zheng Q, Ma Q, Gadidasu KK, Zhang P. Cassava genetic transformation and its application in breeding. Journal of Integrative Plant Biology. 2011;**53**(7): 552-569

[16] Zhang P. Studies on cassava (*Manihot esculenta* Crantz)

transformation: Towards genetic improvement [PhD thesis]. Zurich, Switzerland: Swiss Federal Institute of Technology; 2000

[17] Darkwa K, Olasanmi B, Asiedu R, Asfaw A. Review of empirical and emerging methods and tools for yam (*Dioscorea* spp.) improvement: Status and prospects. Plant Breeding. 2020;**00**:1-24. DOI: 10.1111/PBR.12783

[18] Sarria RA, Torres E, Balcazar N, Destefano-Beltran L, Roca WM. Progress in Agrobacterium-mediated transformation of cassava (*Manihot esculenta* Crantz). In: Cassava Biotechnology Network (ed). Proc. Second Int. Scientific Meeting. Bogor, Indonesia, CIAT Working Document 1995;**150**:241-244

[19] Li HQ, Sautter C, Potrykus I, Puonti-Kaerlas J. Genetic transformation of cassava (Manihot esculenta Crantz). Nature Biotechnology. 1996;**14**:736-740

[20] Gonzalez AE, Schöpke C, Taylor NJ, Beachy RN, Fauquet CM. Regeneration of transgenic cassava plants (*Manihot esculenta* Crantz) through Agrobacterium-mediated transformation of embryogenic suspension cultures. Plant Cell Report. 1998;**17**:827-831

[21] Zhang P, Legris G, Coulin P, Pounti-Kaerlas J. Production of stably transformed cassava plants via particle bombardment. Plant Cell Report. 2000;**19**:939-945

[22] Sarria R, Torres E, Angel F, Chavarriaga P, Roca WM. Transgenic plants of cassava (*Manihot esculenta*) with resistance to Basta obtained by Agrobacterium-mediated transformation. Plant Cell Report. 2000;**19**:339-344

[23] Schreuder MM, Raemakers CJJM, Jacobsen E, Visser RGF. Efficient production of transgenic plants by Agrobacterium-mediated transformation of cassava (*Manihot esculenta* Crantz). Euphytica. 2001;**120**: 35-42

[24] Zhang P, Jaynes J, Potrykus I, Gruissem W, Pounti-Kaerlas J. Transfer and expression of an artificial storage protein (ASP1) gene in cassava (*Manihot esculenta* Crantz). Transgenic Research. 2003;**12**:243-250

[25] Siritunga D, Sayre RT. Generation of cyanogen-free transgenic cassava. Planta. 2003;**217**:367-373

[26] Jørgensen K, Bak S, Busk PK, Sørensen C, Olsen CE, PuontiKaerlas J, et al. Cassava plants with a depleted cyanogenic glucoside content in leaves and tubers. Distribution of cyanogenic glucosides, their site of synthesis and transport, and blockage of the biosynthesis by RNA interference technology. Plant Physiology. 2005;**139**: 363-374

[27] Saelim L, Phansiri S, Suksangpanomrung M, Netrphan S, Narangajavana J. Evaluation of a morphological marker selection and excision system to generate marker-free transgenic cassava plants. Plant Cell Report. 2009;**28**:445-455

[28] Zhang P, Wang WQ, Zhang GL, Kaminek M, Dobrev P, Xu J, et al. Senescence-inducible expression of isopentenyl transferase extends leaf life, increases drought stress resistance and alters cytokinin metabolism in cassava. Journal of Integrative Plant Biology. 2010;**52**:653-669

[29] Zhang P, Bohl-Zenger S, Pounti-Kaerlas J, Potrykus I, Gruissem W. Two cassava promoters related to vascular expression and storage root formation. Planta. 2003;**218**:192-203

[30] Beltran J, Prıas M, Al-Babili S, Ladino Y, Lopez D, Beyer P, et al.

Expression pattern conferred by a glutamic acid-rich protein gene promoter in field-grown transgenic cassava (*Manihot esculenta* Crantz). Planta. 2010;**231**:1413-1424

[31] Vanderschuren H, Alder A, Zhang P, Gruissem W. Dosedependent RNAi-mediated geminivirus resistance in the tropical root crop cassava. Plant Molecular Biology. 2009;**70**:265-272

[32] Abhary M, Siritunga D, Stevens G, Taylor NJ, Fauquet CM. Transgenic biofortification of the starchy staple cassava (*Manihot esculenta*) generates a novel sink for protein. PLoS One. 2011;**6**(1):e16256

[33] Yadav JS, Ogwok E, Wagaba H, Patil BL, Bagewadi B, Alicai T, et al. RNAi-mediated resistance to cassava brown streak Uganda virus in transgenic cassava. Molecular Plant Pathology. 2011;**12**:677-687. DOI: 10.1111/j.1364-3703.2010.00700.x

[34] White FF, Taylor BH, Huffman GA, Gordon MP, Nester EW. Molecular and genetic analysis of the transferred DNA regions of the root-inducing plasmid of *Agrobacterium rhizogenes*. Journal of Bacteriology. 1985;**164**(1):33-44

[35] Aoki S. Resurrection of an ancestral gene: Functional and evolutionary analyses of the Ngrol genes transferred from Agrobacterium to Nicotiana. Journal of Plant Research. 2004;**117**(4): 329-337

[36] Shibata D, Liu YG. Agrobacterium-mediated plant transformation with large DNA fragments. Trends in Plant Science. 2000;**5**:354-357

[37] Quain MD, Egnin M, Bey B, Thompson R, Bonsi C. Transgenic potential of *Dioscorea rotundata* using Agrobacterium-mediated genetic transformation. Aspects of Applied Biology. 2011;**110**:71-79

[38] Schopke C, Taylor N, Cárcamo R, Konan NK, Marmey P, Henshaw GG, et al. Regeneration of transgenic cassava plants (*Manihot esculenta* Crantz) from microbombarded embryogenic suspension cultures. Nature Biotechnology. 1996;**14**:731-735

[39] Munyikwa TRI, Reamakers CCJM, Schreuder M, Kok R, Schippers M, Jacobsen E, et al. Pinpointing towards improved transformation and regeneration of cassava (*Manihot esculenta* Crantz). Plant Science. 1998;**135**:87-101

[40] Zhang P, Puonti-Kaerlas J. PIG-mediated cassava transformation using positive and negative selection. Plant Cell Report. 2000;**19**:1041-1048

[41] Zhang P, Futterer J, Frey P, Potrykus I, Puonti-Kaerlas J, Gruissem W. Engineering virus-induced ACMV resistance by mimicking a hypersensitive reaction in transgenic cassava plants. In: Vasil IK, editor. Plant Biotechnology 2002 and Beyond. Kluwer Academic Publishers: Dordrecht; 2003. pp. 143-146

[42] Zhang P, Gruissem W. Production of Transgenic Cassava (*Manihot esculenta* Crantz). Transgenic Crops of the World— Essential Protocols. Chapter 23. The Netherlands: Kluwer Academic Publishers; 2004. pp. 301-319

[43] Bull SE, Owiti JA, Niklaus M, Beeching JR, Gruissem W, Vanderschuren H. Agrobacterium-mediated transformation of friable embryogenic calli and regeneration of transgenic cassava. Nature Protocols. 2009;**4**:1845-1854

[44] Li HQ, Huang YW, Liang CY, Guo JY, Liu HX, Potrykus I, et al. Regeneration of cassava plants via shoot organogenesis. Plant Cell Report. 1998;**17**:410-414

[45] Zhang P, Phansiri S, Pounti-Kaerlas J. Improvement of

cassava shoot organosgenesis by the use of silver nitrate in vitro. Plant Cell Tissue and Organ Culture. 2001;**67**: 47-54

[46] Baisakh N, Rehana S, Rai M, Oliva N, Tan J, Mackill D, et al. Marker-free transgenic (MFT) nearisogenic introgression lines (NILs) of 'golden' indica rice (cv IR64) with accumulation of provitaminA in the endosperm tissue. Plant Biotechnology Journal. 2006;**4**: 467-475

[47] Dale EC, Ow DW. Intra and inter-molecular site-specific recombination in plant cells mediated by bacteriophage P1 recombinase. Gene. 1990;**91**:79-85

[48] Ebinuma H, Sugita K, Matsunaga E, Yamakado M. Selection of marker-free transgenic plants using the isopenthenyl transferase gene. Proceedings of the National Academy of Sciences of the United States of America. 1990;**94**: 2117-2121

[49] Zuo J, Niu QW, Moller SG, Chua NH. Chemical-regulated, site-specific DNA excision in transgenic plants. Nature Biotechnology. 2001;**19**: 157-161

[50] Penna S, Sagi L, Swennen R. Positive selectable marker genes for routine plant transformation. In Vitro Cellular and Developmental Biology Plant. 2002;**38**:125-128

[51] Stoykova P, Stoeva-Popova P. PMI (manA) as a nonantibiotic selectable marker gene in plant biotechnology. Plant Cell, Tissue and Organ Culture. 2010;**105**:141-148

[52] Haldrup A, Petersen SG, Okkels FT. The xylose isomerase gene from *Thermoanaerobacterium thermosulfurogenes* allows effective selection of transgenic plant cells using D-xylose as the selection agent. Plant Molecular Biology. 1998;**37**:287-296

[53] De Souza CR, Carvalho LJ, de Mattos Cascardo JC. Comparative gene expression study to identify genes possibly related to storage root formation in cassava. Protein and Peptide Letters. 2004;**11**:577-582

[54] De Souza CR, Carvalho LJ, De Almeida ER, Gander ES. A cDNA sequence coding for a glutamic acid-rich protein is differentially expressed in cassava storage roots. Protein and Peptide Letters. 2006;**13**:653-657

[55] De Souza CR, Aragao FJ, Moreira EC, Costa CN, Nascimento SB, Carvalho LJ. Isolation and characterization of the promoter sequence of a cassava gene coding for Pt2L4, a glutamic acid-rich protein differentially expressed in storage roots. Genetics and Molecular Research. 2009;**8**(1):334-344

[56] Zhao SS, Dufour D, Sanchez T, Ceballos H, Zhang P. Development of waxy cassava with different biological and physico-chemical characteristics of starches for industrial applications. Biotechnology and Bioengineering. 2011;**108**:1925-1935

[57] Ihemere U, Arias-Garzon D, Lawrence S, Sayre R. Genetic modification of cassava for enhanced starch production. Plant Biotechnology Journal. 2006;**4**:453-465

[58] Arango J, Salazar B, Welsch R, Sarmiento F, Beyer P, Al-Babili S. Putative storage root specific promoters from cassava and yam: Cloning and evaluation in transgenic carrots as a model system. Plant Cell Reports. 2010;**29**:651-659

[59] Legg JP, Fauquet CM. Cassava mosaic geminiviruses in Africa. Plant Molecular Biology. 2004;**56**:585-599

[60] Patil BL, Fauquet CM. Cassava mosaic geminiviruses: Actual

knowledge and perspectives. Molecular Plant Pathology. 2009;**10**:685-701

[61] Vanderschuren H, Stupak M, Futterer J, Gruissem W, Zhang P. Engineering resistance to geminiviruses—Review and perspectives. Plant Biotechnology Journal. 2007;**5**:207-220

[62] Chellappan P, Masona MV, Vanitharani R, Taylor NJ, Fauquet CM. Broad spectrum resistance to ssDNA viruses associated with transgene-induced gene silencing in cassava. Plant Molecular Biology. 2004;**56**:601-611

[63] Zhang P, Vanderschuren H, Futterer J, Gruissem W. Resistance to cassava mosaic disease in transgenic cassava expressing antisense RNAs targeting virus replication genes. Plant Biotechnology Journal. 2005;**3**:385-397

[64] Vanderschuren H, Akbergenov R, Pooggin MM, Hohn T, Gruissem W, Zhang P. Transgenic cassava resistance to African cassava mosaic virus is enhanced by viral DNA-A bidirectional promoter-derived siRNAs. Plant Molecular Biology. 2007;**64**:549-557

[65] Bi H, Aileni M, Zhang P. Evaluation of cassava varieties for cassava mosaic disease resistance jointly by agro-inoculation screening and molecular markers. African Journal of Plant Science. 2010;**4**(9):330-338

[66] Mbanzibwa DR, Tian Y, Mukasa SB, Valkonen JPT. Cassava brown streak virus (potyviridae) encodes a putative Maf/HAM1 pyrophosphatase implicated in reduction of mutations and a P1 proteinase that suppresses RNA silencing but contains no HC-Pro. Journal of Virology. 2009;**83**:6934-6940

[67] Winter S, Koerbler M, Stein B, Pietruszka A, Paape M, Butgereitt A. Analysis of cassava brown streak viruses reveals the presence of distinct virus species causing cassava brown streak disease in East Africa. Journal of General Virology. 2010;**91**:1365-1372

[68] Patil BL, Ogwok E, Wagaba H, Mohammed IU, Yadav JS, Bagewadi B, et al. RNAi-mediated resistance to diverse isolates belonging to two virus species involved in cassava brown streak disease. Molecular Plant Pathology. 2011;**12**(1):31-41

[69] Lopez C, Soto M, Restrepo S, Piegu B, Cooke R, Delseny M, et al. Gene expression profile in response to *Xanthomonas axonopodis* pv. Manihotis infection in cassava using a cDNA microarray. Plant Molecular Biology. 2005;**57**:393-410

[70] Tripathi JN, Ntui VO, Ron M, Muiruri SK, Britt A, Tripathi L. CRISPR/Cas9 editing of endogenous banana streak virus in the B genome of *Musa* spp. overcomes a major challenge in banana breeding. Communications Biology. 2019;**2**(46):1-11

[71] Seal S, Turaki A, Muller E, Kumar PL, Kenyon L, Filloux D, et al. The prevalence of badnaviruses in West African yams (*Dioscorea cayenensis-rotundata*) and evidence of endogenous pararetrovirus sequences in their genomes. Virus Research. 2014;**186**:144-154. DOI: 10.1016/j.virus res.2014.01.007

[72] Umber M, Filloux D, Muller E, Laboureau N, Galzi S, Roumagnac P, et al. The genome of African yam (Dioscorea cayenensis-rotundata complex) hosts endogenous sequences from four distinct badnavirus species. Molecular Plant Pathology. 2014;**15**(8):90-801

[73] Roy B, Zhao J, Yang C, Luo W, Xiong T, Li Y, et al. CRISPR/Cascade 9-mediated genome editing-challenges and opportunities. Frontier in Genetics. 2018;**9**(240):1-12. DOI: 10.3389/fgene.2018.00240

[74] Kleinstiver BP, Pattanayak V, Prew MS, Tsai SQ, Nguyen NT, Zheng Z, et al. High-fidelity CRISPR–Cas9 nucleases with no detectable genome-wide off-target effects. Nature. 2016;**529**:490-495. DOI: 10.1038/nature16526

[75] Zetsche B, Gootenberg JS, Abudayyeh OO, Slaymaker IM, Makarova KS, Essletzbichler P, et al. Zhang F Cpf1 is a single-RNA-guided endonuclease of a Class 2 CRISPR-Cas system. Transgenic Research. 2016;**25**: 207

[76] Feng S, Song W, Fu R, Zhang H, Xu A, Li J. Application of the CRISPR/Cas9 system in *Dioscorea zingiberensis*. Plant Cell, Tissue and Organ Culture (PCTOC). 2018;**135**(1):133-141. DOI: 10.1007/s11240-018-1450-5

[77] Tripathi L, Syombua E, Tripathi J, Ntui V, Wang K, Yang B. *Prospects of Genetic Engineering and Gene Editing in Yam Improvement*. Paper Presented at the 17th Plant and Animal Genome Conference; 12th-16th January 2019; San Diego, USA

[78] Sakurai T, Planta G, Rodríguez-Zapata F, Seki M, Salcedo A, Toyoda A, et al. Sequencing analysis of 20000 full-length cDNA clones from cassava reveals lineage specific expansions in gene families related to stress response. BMC Plant Biology. 2007;7:66-104

[79] Li K, Zhu W, Zeng K, Zhang Z, Ye J, Ou W, et al. Proteome characterization of cassava (*Manihot esculenta* Crantz) somatic embryos, plantlets and tuberous roots. Proteome Science. 2010;**8**:10

[80] Mitprasat M, Roytrakul S, Jiemsup S, Boonseng O, Yokthongwattana K. Leaf proteomic analysis in cassava (*Manihot esculenta* Crantz) during plant development, from planting of stem cutting to storage root formation. Planta. 2011;**233**:1209-1221. DOI: 10.1007/s00425-011-1373-4

[81] Stupak M, Vanderschuren H, Gruissem W, Zhang P. Biotechnological approaches to cassava protein improvement. Trends Food Science and Technology. 2006;**17**:634-641

[82] Vimala B, Thushara R, Nambisan B, Sreekumar J. Effect of processing on the retention of carotenoids in yellow-fleshed cassava (*Manihot esculenta* Crantz) roots. International Journal of Food Science and Technology. 2011;**46**: 166-169

[83] White WLB, Arias-Garzon DI, McMahon JM, Sayre RT. Cayanogenesis in cassava: The role of hydroxynitrile lyase in root cyanide production. Plant Physiology. 1998;**116**:1219-1225

[84] Siritunga D, Sayre RT. Engineering cyanogen synthesis and turnover in cassava (*Manihot esculenta*). Plant Molecular Biology. 2004;**56**:661-669

[85] Munyikwa TRI, Langeveld S, Salehuzzaman SNIM, Jacobsen E, Visser RGF. Cassava starch biosynthesis: New avenues for modifying starch quantity and quality. Euphytica. 1997;**96**:65-75

[86] Baguma Y, Sun C, Ahlandsberg S, Mutisya J, Palmqvist S, Rubaihayo PR, et al. Expression patterns of the gene encoding starch branching enzyme II in the storage roots of cassava (*Manihot esculenta* Crantz). Plant Science. 2003;**164**:833-839

[87] Baguma Y, Sun C, Boren M, Olsson H, Rosenquist S, Mutisyá J, et al. Sugar-mediated semidian oscillation of gene expression in the cassava storage root regulates starch synthesis. Plant Signal Behavior. 2008;**3**:439-445

[88] Jobling SA, Westcott RJ, Tayal A, Jeffcoat R, Schwall GP. Production of a freeze–thaw-stable potato starch by

antisense inhibition of three starch synthase genes. Nature Biotechnology. 2002;**20**:295-299

[89] Raemakers K, Schreuder M, Suurs L, Furrer-Verhorst H, Vincken JP, de Vetten N, et al. Improved cassava starch by antisense inhibition of granule-bound starch synthase I. Molecular Breeding. 2005;**16**:163-172

[90] Ma QX, Xu J, Qiao AM, Zhang P. Current progress in studies on post-harvest physiological deterioration of cassava storage roots. Journal of Tropical Subtropical Botany. 2009;**17**(3):309-314

[91] Reilly K, Bernal D, Cortes DF, Gomez-Vasquez R, Tohme J, Beeching JR. Towards identifying the full set of genes expressed during cassava post-harvest physiological deterioration. Plant Molecular Biology. 2007;**64**:187-203

[92] Blagbrough IS, Bayoumi SAL, Rowan MG, Beeching JR. Cassava: An appraisal of its phytochemistry and its biotechnological prospects. Phytochemistry. 2010;**71**:1940-1951

[93] Morante N, Sanchez T, Ceballos H, Calle F, Perez JC, Egesi C, et al. Tolerance to postharvest physiological deterioration in cassava roots. Crop Science. 2010;**50**:1333-1338

[94] Abang MM, Winter S, Green KR, Hoffmann P, Mignouna HD, Wolf GA. Molecular identification of Colletotrichum gloeosporioides causing anthracnose of yam in Nigeria. Plant Pathology. 2002;**51**:63-71. DOI: 10.1046/j.0032-0862.2001.00655.x

[95] Keller H, Pamboukdjian N, Ponchet M, Poupet A, Delon R, Verrier JL, et al. Pathogen-induced elicitin production in transgenic tobacco generates a hypersensitive response and nonspecific disease resistance. The Plant Cell. 1999;**11**(2):223-235. DOI: 10.1105/tpc.11.2.223

[96] Lorito M, Woo SL, Garcia I, Colucci G, Harman GE, Pintor-Toro JA, et al. Genes from mycoparasitic fungi as a source for improving plant resistance to fungal pathogens. Proceedings of the National Academy of Sciences of the United States of America. 1998;**95**:7860-7865. DOI: 10.1073/pnas.95.14.7860

[97] Broekaert WF, Cammue BPA, De Bolle MFC, Thevissen K, De Samblanx GW, Osborn RW. Antimicrobial peptides from plants. Critical Review of Plant Science. 1997;**16**:297-323. DOI: 10.1080/07352689709701952

[98] Atkinson HJ, Urwin PE, McPherson MJ. Engineering plants for nematode resistance. Annual Review of Phytopathology. 2003;**41**:615-639. DOI: 10.1146/annurev.phyto.41.052002.095737

[99] Urwin PE, Lilley CJ, McPherson MJ, Atkinson HJ. Resistance to both cyst and root-knot nematodes conferred by transgenic Arabidopsis expressing a modified plant cystatin. Plant Journal. 1997;**12**:455-461. DOI: 10.1046/j.1365-313X.1997.12020455.x

[100] Atkinson HJ, Urwin PE, Clarke MC, McPherson MJ. Image analysis of the growth of *Globodera pallida* and *Meloidogyne incognita* on transgenic tomato roots expressing cystatins. Journal of Nematology. 1996;**28**:209-215

[101] Vain P, Worland B, Clarke MC, Richard G, Beavis M, Liu H, et al. Expression of an engineered cysteine proteinase inhibitor (oryzacystatin-I_D86) for nematode resistance in transgenic rice plants. Theoretical Applied Genetics. 1998;**96**:266-271. DOI: 10.1007/s001220050735

[102] Urwin PE, Troth KM, Zubko EI, Atkinson HJ. Effective transgenic resistance to *Globodera pallida* in potato field trials. Molecular Breeding.

2001;**8**:95-101. DOI: 10.1023/A:1011942003994

[103] Chan YL, Yang AH, Chen JT, Yeh KW, Chan MT. Heterologous expression of taro cystatin protects transgenic tomato against Meloidogyne incognita infection by means of interfering sex determination and suppressing gall formation. Plant Cell Reports. 2010;**29**:231-238. DOI: 10.1007/s00299-009-0815-y

[104] Gao S, Yu B, Zhai H, He SZ, Liu QC. Enhanced stem nematode resistance of transgenic sweetpotato plants expressing oryzacystatin-I gene. Agricultural Sciences in China. 2011;**10**(4):519-525. DOI: 10.1016/S1671-2927(11)60032-1

[105] Roderick H, Tripathi L, Babirye A, Wang D, Tripathi JN, Urwin PE, et al. Generation of transgenic plantain (*Musa* spp.) with resistance to plant pathogenic nematodes. Molecular Plant Pathology. 2012;**13**:842-851. DOI: 10.1111/j.1364-3703.2012.00792.x

[106] Kim SH, Ahn YO, Ahn MJ, Lee HS, Kwak SS. Down-regulation of β-carotene hydroxylase increases β-carotene and total carotenoids enhancing salt stress tolerance in transgenic cultured cells of sweetpotato. Phytochemistry. 2012;**74**:69-78

[107] Kim SH, Jeong JC, Park S, Bae JY, Ahn MJ, Lee HS, et al. Down-regulation of sweetpotato lycopene β-cyclase gene enhances tolerance to abiotic stress in transgenic calli. Molecular Biology Reports. 2014;**41**:8137-8148

[108] Park SC, Kim SH, Park S, Lee HU, Lee JS, Park WS, et al. Enhanced accumulation of carotenoids in sweetpotato plants overexpressing IbOr-Ins gene in purple-fleshed sweetpotato cultivar. Plant Physiology and Biochemistry. 2015;**86**:82-90

[109] Mano H, Ogasawara F, Sato K, Higo H, Minobe Y. Isolation of a regulatory gene of anthocyanin biosynthesis in tuberous roots of purple-fleshed sweet potato. Plant Physiology. 2007;**143**:1252-1268

[110] Park SC, Kim YH, Kim SH, Jeong YJ, Kim CY, Lee JS, et al. Overexpression of the IbMYB1 gene in an orange-fleshed sweet potato cultivar produces a dual-pigmented transgenic sweet potato with improved antioxidant activity. Physiologia Plantarum. 2015;**153**(4):525-537

[111] Tanaka M, Takahata Y, Kurata R, Nakayama H, Yoshinaga M. Structural and functional characterization of IbMYB1 genes in recent Japanese purple-fleshed sweetpotato cultivars. Molecular Breeding. 2012;**29**:565-574

[112] Padda MS, Picha DH. Quantification of phenolic acids and antioxidant activity in sweetpotato genotypes. Scientia Horticulturae. 2008;**119**:17-20

Chapter 3

Next Generation of Transgenic Plants: From Farming to Pharming

Alp Ayan, Sinan Meriç, Tamer Gümüş and Çimen Atak

Abstract

The number of approaches related to recombinant protein production in plants is increasing rapidly day by day. Plant-based expression offers a safe, cost-effective, scalable, and potentially limitless way to rapidly produce recombinant proteins. Plant systems, which have significant advantages over animal and yeast recombinant protein production systems, are particularly promising for the large-scale production of antibodies and therapeutic proteins. Molecular pharming with transgenic plant systems become prominent among other production systems with its low cost, absence of human or animal pathogen contaminants, and the ability to use post-translational modifications such as glycosylation. The ability to produce recombinant pharmaceutical proteins in plant seeds, plant cells and various plant tissues such as hairy roots and leaves, through the stable transformation of the nuclear genome or transient expression, allows for the establishment of different production strategies. In particular, the rapid production of candidate proteins by transient expression, which eliminates the need for lengthy transformation and regeneration procedures, has made plants an attractive bioreactor for the production of pharmaceutical components. This chapter aimsto exhibit the current plant biotechnology applications and transgenic strategies used for the production of recombinant antibodies, antigens, therapeutic proteins and enzymes, which are used especially in the treatment of various diseases.

Keywords: molecular pharming, plant-derived pharmaceutical, therapeutic proteins, transient expression, recombinant proteins

1. Introduction

In the last couple of decades, many initiatives have been carried out in which different disciplines came together for sustainable farming in respect to the increasing food demands of the world population which is expected to reach 9 billion thresholds in 2050. Conventional breeding, mutation breeding and especially transgenic technology have been frequently used to prevent yield losses which are caused by the decreasing amount of arable land due to various reasons such as urbanization, desertification and salinity and drought, various diseases, weeds and insects. With advances in modern biotechnology, many approaches have been used to improve crop varieties, from marker-assisted selection (MAS) to recently developed new plant breeding techniques (NPBTs). In particular, genetic modifications have significantly expanded the genetic pool that has been used by plant breeders since the mid-90s. Thus, the development of new plants with many different agricultural traits has gained great momentum.

IntechOpen

After the acceleration of commercialization of transgenic crops, which were initially developed for agronomic purposes such as insect and herbicide tolerance for only producers, especially stacked GM events, in which two or more characteristics are introduced together, emerged. Approximately 30 different transgenic plants with many different characteristics for both the producer and the consumer as disease resistance, abiotic stress tolerance, increase in nutrition and food quality, including fruits and vegetables as *Phaseolus vulgaris* (bean), *Solanum melongen* (eggplant), *Cucumis melo* (melon), *Carica papaya* (papaya), *Prunus domestica* (plum), *Beta vulgaris* (sugar beet), had been approved [1, 2]. Stacked GM crops have a combination of several traits and even four different traits (genetic modification) are possible to be located in one GM event, as in the Widestrike™ Roundup Ready Flex™ cotton and Herculex™ RW Roundup Ready™-2 maize samples [3]. Moreover, in the last decade, new genome editing techniques such as zinc finger nuclease (ZFN) technology, clustered regularly interspaced short palindromic repeat (CRISPR)-associated nuclease systems and transcription activator-like effector nucleases (TALENs), which provides precise genome modifications, enabled the reproduction of plants which does not contain recombinant DNA. The recently adopted cisgenesis/intragenesis, RNA interference and new genome editing techniques have enabled the development of many genetically edited organisms (GEOs) as well.

Along with all these developments, it has been understood after the 2000s that GMOs, which have become increasingly dominant in the agricultural field, can also be utilized effectively in medicine and industrial fields. Especially, the increase in chronic and infectious diseases as a result of overpopulation, and the outbreak of epidemics and pandemics cause demands that exceed the production capacity of molecules used for diagnosis, treatment and prevention. To satisfy these demands, cost-effective systems that can produce high-quality products, and allow to take action quickly are developed. In addition, it is a fact that models that support small-scale production will be needed in recent years, as is the case with personalized medicines. Especially in the last two decades, it has been possible to produce plant-based pharmaceutical proteins that can be used in the diagnosis and treatment of many diseases with the developments in gene transfer and production strategies of recombinant proteins in plants.

Thanks to improvements in basic points such as new transformation strategies (stable or transient), transformation methods, appropriate promoter selection and codon optimization, high-scale and lower cost recombinant proteins in plants can be produced rapidly [4, 5]. Thus, the interest in producing recombinant proteins for direct use as a product without aiming for a specific change in phenotype or metabolism and obtaining them in purified form or crude plant extract has increased. This potential provided by the production of recombinant pharmaceutical proteins in plants for therapeutic applications was realized much later than in bacterial, yeast and animal systems. On the other hand, in the early periods, interest in the production of non-pharmaceutical products such as industrial enzymes, cosmetic ingredients, biosensors or biocatalysts, feed, biofuel in plants due to their short development times, low purification costs and less regulatory burdens was much higher than pharmaceutical products [6].

The first applications of molecular farming were come to fruition in the late 1980s by producing various antibodies and human-specific proteins in transgenic plants. Among all expression systems, plant-based systems have started to be preferred as an attractive alternative with their advantages in low production cost, high level of transgene expression, rapid scalability, the riskless transmission of human and animal pathogens and production of proteins with secondary modification (**Table 1**). It is also possible to produce recombinant proteins in dry tissues such as seeds or grains, which reduces storage and transportation costs and puts plant

DOI: http://dx.doi.org/10.5772/intechopen.102004

	Production cost	High scale production	Development times	Product quality	Glycosylation	Purification-storage cost and diffuculty	Safety
Bacteria	Low	High	Low	Low	None	Low	Medium (Endotoxi)
Yeast	Medium	High	Medium	Medium	Incorrect	Low	Medium
Insect	Medium	Medium	Medium	Medium	Correct	Medium	Low
Mammalian cel culture	High	Low	High	Very High	Correct	Medium	Very Low (Virus, prion and oncogens)
Transgenic Animals	Very high	Medium	Very high	Very High	Correct	Medium	Very Low (Virus, prion and oncogens)
In vitro plant culture (Hairy roots, cell suspension culture)	Medium	High	High	High	Minor differences	Medium	Very High
Stable transgenic plants	Medium	Medium	Very High	High	Minor differences	High*	Very High
Transient plants	Low	High	Low	High	Minor differences	High*	Very High

Edible forms don't have any purification-storage cost and difficulty.

Table 1.
Comparison of different production systems according to their efficiency.

systems one step ahead. While plants can produce complex proteins similar to other eukaryotic systems, they can modify these proteins post-translationally. However, plants can present enhanced post-translational glycosylation modifications, unlike yeast and insects, which have prokaryotic expression systems or very simple glycosylation patterns. Plant cells have slight differences in their glycosylation patterns compared to mammalian cells, and these differences can be rearranged by genetic engineering [7]. Since different glycan structures affect the stability of glycoproteins, subcellular targeting, immunogenicity, pharmacokinetic behavior and biological activity, it is essential to produce the recombinant protein with the appropriate glycosylation pattern [8, 9]. One of the most significant strategies which are developed to control the glycosylation of recombinant proteins in plants is subcellular targeting that prevent the addition of undesired sugar residues. Another important strategy is glycoengineering that avoids the addition of plant glycans and even replaces them with human proteins [9]. In the following years, studies have accelerated and the term of molecular pharming has become prominent with the increasing demand for the production of recombinant pharmaceutical proteins in many different plant systems [9].

Following the enhanced knowledge of gene transfer to plants and finely-tuned gene transfer methods specific to many species such as *Nicotiana tabacum* and *N. benthamiana* (tobacco plant), *Lactuca sativa* (lettuce), *Glycine max* (soybean), *Oryza sativa* (rice), *Solanum lycopersicum* (tomato), *Medicago sativa* (alfalfa), *Zea mays* (maize), it is possible to produce many recombinant proteins including monoclonal antibodies, enzymes, growth factors, therapeutic proteins and vaccines in various plant tissues or cell. Recombinant protein production can be accomplished in the whole plant, as well as in certain parts such as seeds, leaves and fruits [10]. The selected production strategy must express the protein with high efficiency and fully come up with the requirements of the regulations in terms of safety and production quality. To produce a recombinant protein, optimization of the coding sequences of a gene that is desired to transfer, determination of the gene expression strategy as stable or transient, selection of convenient plants and cost-efficient methods for isolation and purification of target proteins are required. Stable gene transfer which is performed by introducing the target gene into the nuclear or plastid genome by *Agrobacterium* and particle bombardment respectively results in stable expression in plant tissues. Although the methods for plant transformation vary according to target species, the target genome (nuclear or plastid), the structure of the gene to be transferred, there are two common approaches: direct (e.g. biolistic or microparticle bombardment) or indirect (*Agrobacterium tumefaciens* or *Agrobacterium rhizogenes*). It is possible to obtain a whole transgenic plant by using plant tissue culture methods together with mentioned transformation methods [11]. Moreover, high-scale production of recombinant proteins becomes possible for many different plant species by establishing various cultures (e.g., callus culture, hairy root culture or suspension culture) from transformed plant tissues or cells [12].

It is crucial to state that developing transgenic plants by stable transformation is more time-consuming by comparing to transient expression systems. In addition, stable transformation strategies need improvement to increase the level of transgene expression. Especially, due to the random insertion of the transgene into the nuclear genome, the different profiles that emerge as a result of the positional effect should be followed meticulously. Unstable gene expression, as well as transgene-induced gene silencing, may occur together with multiple insertions [4]. Unlike stable expression, when we focus on recombinant protein production strategies based on transient expression of transgenes carried by bacteria (*A. tumefaciens*) or viral vectors (tobacco mosaic virus (TMV), cauliflower mosaic virus (CMV), alfalfa mosaic virus (AVM)), it is seen obviously that stable integration of transgene is not necessary. Thus, it becomes possible to produce recombinant protein rapidly by avoiding time-consuming transformation and regeneration procedures. With transient expression by *Agrobacterium* infiltration or viral vectors, it is possible to ensure rapid expression of the transgene even 3–4 hours after transformation and to produce recombinant protein by reaching the maximum expression level in the range of 18–48 hours [4]. Other important advantages of transient expression are that they can reach higher expression levels compared to stable transformation and maintain gene expression in the range of 10–14 days.

One of the important handicaps in recombinant protein production in plants is the lack of suitable, reliable and inexpensive purification methods for each plant. This deadlock becomes even more difficult when the purification of pharmaceuticals must fulfill the stringent criteria mandated by the "good manufacturing practice (GMP)" standards. It is known that purification practices performed in accordance with legal standards comprise approximately 80% of the total costs [13]. Apart from purification by chromatography, membrane filtration and fusion of various polypeptides, there is a need to develop more efficient and cost-effective purification strategies. On the other hand, with the redesign of plant parts containing the target recombinant protein in a form suitable for an oral route such as edible vaccines, it is possible to get rid of the heavy financial burdens of purification strategies. One of the issues that need to be carefully considered regarding the production of pharmaceutical proteins in plants is regulation and biosafety issues. Especially, economic damages to farmers and the food industry as a result of the co-mingling of food and feed crops with plants designed for pharmaceutical production, the spread of transgenes through pollen or seeds, undesirable exposure of non-target organisms such as insects, birds and horizontal gene transfer are among the emerging risks [14].

2. Transgenic plants in farming

Since the discovery of genetic mechanisms of reproduction and biodiversity, plant breeders methodically try to exploit agronomically desired traits for more profitable crop production in many aspects. In the last quarter of the twentieth century, genome manipulation techniques known as genetic engineering were introduced into various organisms. In the early years of this novel technology in plant science, desired traits were related to higher yield, resistance to various biotic and abiotic stress factors. Considerable success has been obtained in enhancing photosynthetic capacity, increasing root and leaf size, stimulating vegetative growth, improving biomass and more. Following transgenic approaches aimed nutritional quality through enhancing various biomolecule production and increased shelf life as well [15]. In recent days, transgenic technology may even offer some solutions to the global energy crisis through improvements in biofuel production. A total of 17 million farmers from 29 different countries surged in biotech crops in the period from 1996 to 2019 after the first successful commercialized release. During these 24 years, global biotech crop planting increased 112- fold from 1.7 million to 190.4 hectares. In the second third of this period, trends in leading biotech crop-producing countries started to change as well. Since 2011, 24 developing countries produced 56% of total biotech crops while 5 leading industrial countries share the rest of the 44%. This trend tends to accelerate as long as other developing countries realize the potential. A total of 19 countries among 29 are considered as "biotech mega" with at least 50,000 hectares planting. United States, Brazil, Argentina, Canada and India are the leading planters. Soybean, maize, cotton and canola are the most planted biotech crops worldwide (**Figure 1**). Plants as alfalfa, sugar beet,

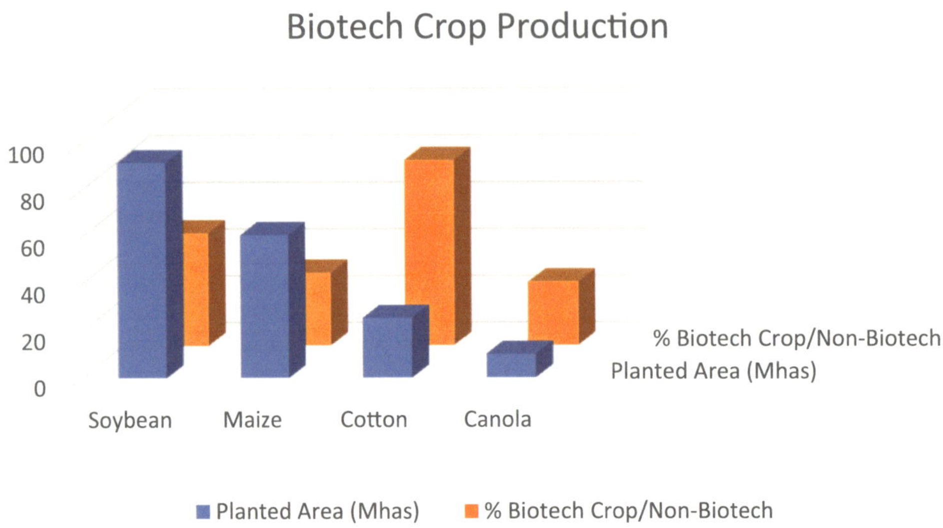

Figure 1.
Worldwide biotech crop production [16].

potato, apples, squash, papaya and eggplant share only 1.8% of total production besides these four plants.

Biotech crops, which have reached 224.9 billion US dollar global value have also contributed to conserving biodiversity by preventing deforestation, reducing pesticide requirement, decreasing CO_2 emissions in agricultural applications and alleviating socio-economic conditions of small farmers.

Transgenic plants are generally classified into three generations. First-generation GM crops were developed against various biotic and abiotic stress factors, while the second generation targeted better nutritional quality [15]. The remarkable amount of progress has been achieved in countless laboratory practices with many different plant species. On the other hand, commercial GM varieties are considerably limited compared to these laboratory practices. As it can be seen in **Table 2**, all commercial GM traits belong to the first and second generations of transgenic approaches. In laboratory practices, abiotic stress tolerance is one of the key aspects of desired crops. However, abiotic stress tolerance mechanisms and pathways are extremely intricate. Defining a particular gene target for an abiotic stress factor and introducing it to a susceptible variety is usually an insufficient strategy due to these complex responses within and between species. Plants have developed stress signal perception and transduction pathways that regulate stress-inducible genes through transcription factors (NAC, WRKY, MYB, bZIP, DREB/CBF), kinases and phosphatases. Main stress-inducible genes are kinases, molecular chaperones, osmoprotectants, transcription factors [17–20].

Many biomolecules are identified with their known protective roles against abiotic stresses. Therefore, these molecules are potential gene targets for transgenic abiotic stress tolerance approaches. Transgenic regulations of solutes such as glycine betaine, mannitol, trehalose and proline which acts as an osmoprotectant, metal chelator, antioxidative defense molecule and signal molecule have been used to enhance stress tolerance in many plants. For an instance, codA expressing GM indica rice plant which has enhanced glycine betaine production through increased choline oxidase activity, present induced water stress tolerance [21]. Abiotic stress factors usually cause misfolding and precipitation of crucial proteins. Heat shock proteins (HSPs) act as molecular chaperones and mediate folding, assembly, translocation and degradation of misfolded proteins. As a molecular chaperon,

DOI: http://dx.doi.org/10.5772/intechopen.102004

Plant / Commercia GM Trait	Abiotic Stress Tolerance	Altered Growth/Yield	Disease Resistance	Herbicide Tolerance	Insect Resistance	Modified Product Quality	Pollination control system	TOTAL
Alfalfa – *Medicago sativa*				4		2		6
Apple – *Malus* x *Domestica*						9		9
Argentine Canola – *Brassica napus*				35			22	57
Bean – *Phaseolus vulgaris*			1					1
Carnation – *Dianthus caryophyllus*				4		1		5
Chicory – *Cichorium intybus*				3			3	6
Cotton – *Gossypium hirsutum* L.				45	50			95
Cowpea – *Vigna unguiculata*					1			1
Creeping Bentgrass – *Agrostis stolonifera*				1				1
Eggplant – *Solanum melongena*					1			1
Eucalyptus – *Eucalyptus sp*.		1				14		15
Flax – *Linum usitatissimum* L.				1				1
Maize – *Zea mays* L.	7	2		215	210		6	440
Melon – *Cucumis melo*								0
Papaya – *Carica papaya*			4					4
Petunia – *Petunia hybrida*						18		18

Plant / Commercia GM Trait	**Abiotic Stress Tolerance**	**Altered Growth/Yield**	**Disease Resistance**	**Herbicide Tolerance**	**Insect Resistance**	**Modified Product Quality**	**Pollination control system**	**TOTAL**
Pineapple – *Ananas comosus*						2		2
Plum – *Prunus domestica*			1			12		13
Polish canola – *Brassica rapa*				4				4
Poplar – *Populus sp.*					2			2
Potato – *Solanum tuberosum* L.			19	4	30			53
Rice – *Oryza sativa* L.				3	3			6
Rose – *Rosa hybrida*						1		1
Safflower – *Carthamus tinctorius* L.						9		9
Soybean – *Glycine max L.*	2	1		35	6			44
Squash – *Cucurbita pepo*			2			3		5
Sugar Beet – *Beta vulgaris*				3		19		22
Sugarcane – *Saccharum spp.*	3				3	2		8
Sweet pepper – *Capsicum annuum*			1			2		3
Tobacco – *Nicotiana tabacum* L.				1		1		2
Tomato – *Lycopersicon esculentum*			1		1	2		4
Wheat – *Triticum aestivum*				1		2		3
Total	12	4	29	359	307	99	31	841

Table 2.
Registered commercial GM traits and related crops [16].

transgenic *Trichoderma harzianum* hsp70 is shown to increase heat and other abiotic stress resistance in *Arabidopsis thaliana* plants [22]. Also, late embryogenesis abundant (LEA) proteins, aquaporins and calcineurin B-like proteins with antioxidant, membrane protection and ion binding functions are widely preferred targets [1]. Rab28 LEA gene over-expressing maize plants are reported to have improved desiccation tolerance under a constitutive maize promoter [23]. Likewise, transgenic expression of halophilic fungus *Aspergillus glaucus* AgGlpF gene encoding an aquaporin protein confers extreme salt tolerance in soybean [24].

Other commercially registered GM traits are altered insect and disease resistance, growth/yield, herbicide tolerance, modified product quality and pollination control system. Insect and disease resistance is mainly obtained through introducing natural pest genes from insects, fungi and bacteria to target plants. The most known example of the application is δ-endotoxin insecticidal protein-expressing cry gene transfer from *Bacillus thuringiensis* (Bt) which is extremely effective against lepidopterans, dipterans and coleopterans. In recent days, there are several alternatives to Bt toxins including lectins, protease inhibitors, antibodies, peptide hormones [15]. In particular, protease inhibitors (PIs) have devastating effects on insect digestive systems since most insects facilitate serine-type proteinase enzymes in digestion. Serine-type mustard trypsin inhibitor −2 (MTI-2) expressing *Brassica napus (L.)* plant are proven to present insecticidal properties on *Pterostichus madidus* beetle. It was also found effective on *Plutella diamondback* moth xylostella as the intermediary pest species [25]. As an alternative to Bt toxin alone, lectin genes were also introduced for insecticidal properties. Codon-optimized synthetic Bt Vip3Aa gene under CaMV35S promoter and *Allium sativum* leaf agglutinin gene under phloem-specific promoter transformation significantly improved *Helicoverpa armigera* resistance. As mentioned in environmental stress factors earlier, transcription factors are known to play important roles in plant resistance to environmental stress factors. MYB4L transcription factors were shown to induce ethylene pathway and enhance tobacco mosaic virus (TMV) resistance in *Nicotiana benthamiana*. On the other hand, silencing the transcription factor intensified the susceptibility [26]. Similarly, the WRKY17 transcription factor was proven to both enhance artemisinin biosynthesis in a traditional Chinese medicinal plant *Artemisia annua,* and provide resistance against *Pseudomonas syringae*. Two defense marker genes, pathogenesis-related 5 (PR5) and NDR1/HIN1-LIKE 10 (NHL10), were significantly increased in AaWRKY17-overexpressing transgenic *A. annua* plants as well as AaWRKY17 directly bound to the promoter region of the artemisinin biosynthetic pathway gene amorpha-4,11-diene synthase (ADS) and promoted its expression.

Herbicide resistance can be maintained through two transgenic approaches. The first involves the modification of the target enzyme to overcome herbicide sensitivity. In the second approach herbicide, detoxifying pathways are introduced to the susceptible target plants. Glyphosate (N-phosphonomethylglycine) is a highly efficient, low-toxicity, broad-spectrum and nonselective herbicide that has been widely applied. Glyphosate specifically inhibits the activity of 5-enolpyruvylshikimate-3-phosphate synthase (EPSPS), a key enzyme in the shikimate pathway and prevents aromatic amino acid synthesis unselectively in any plants. Therefore, fortification of target enzymes in intended plants is crucial. Recently, co-transformation of a codon-optimized glyphosate oxidase gene WBceGO-B3S1 from a variant BceGO-B3S1 and a glyphosate-tolerant gene I. variabilis-EPSPS from the bacterium *Isoptericola variabilis* into an *O. sativa* variety by *Agrobacterium*-mediated genetic transformation resulted in high glyphosate tolerance [27]. In the second approach, the herbicidal effects of glufosinate, which is a nonselective, glutamine synthetase (GS) inhibitor, are targeted. The GS enzyme produces glutamine amino acids from ammonia and glutamate. Glufosinate causes glutamine deficiency in susceptible plants through ammonia and glyoxylate accumulation, inhibition of photosynthesis due to defected chloroplast structure. Transformation of the pat gene from *Streptomyces viridochromogenes* to susceptible plants leads to expression of phosphinothricin acetyltransferase (PAT) enzyme in plants which metabolize glufosinate into N-acetyl-L-glufosinate (NAG). This non-toxic compound does not inhibit GS enzymes [28].

The first two generations of transgenic plants in farming applications consistently enhanced food supply and essential traits. Following the improvements within this technology, it became more friendly to the environment, less risky for human health, more profitable to small scale farmers and more regulated around the world. The third generation will be the era of recombinant pharmaceuticals as plant-based vaccines against pathogens in human health and beneficial health products as therapeutic proteins, monoclonal antibodies, hormones, enzymes along with transgenic bioenergy plants which are not going to be discussed in this chapter.

3. Transgenic plants in pharming

Excluding transgenic farming applications which were summarized in the previous section, the term of molecular farming through transgenic plants refers utilization of plants for recombinant protein production instead of microorganisms and animal cell in fermenter systems. When we narrow this definition from recombinant proteins to pharmaceutical products still in plants basis, we define the alternative term "molecular pharming". Before the revolutionary contribution of Stieger et al. [29] in which they presented the capability of plant cells to produce multimeric assembled mammalian antibodies in *Acetabularia mediterranea*, the general opinion was that functional full-length mammalian antibodies would not assemble in plants without mammalian chaperones. Before this accomplishment, there were various well-established production systems as *Escherichia coli* in prokaryotes, *Saccharomyces cerevisiae* in single-cell eukaryotes, Chinese Hamster Ovary (CHO) cells, non-secreting murine myeloma (NS0) cells, Sp2/0 HEK293 mammalian cells for recombinant protein production in which there were boundaries of the expensive fermenter and good manufacturing practice (GMP) required for pharmaceutical production. After successful production practices of mammalian IgG1 antibody in tobacco [30] and human serum albumin (HSA) in transgenic potato and tobacco plants [31], the molecular pharming approach rapidly accelerated and production systems like tobacco, potato, tomato, alfalfa, safflower, carrot, lettuce, strawberry, moss, duckweed, maize, wheat and rice were emerged. Besides the plant species, there are various approaches differing in plant tissues (whole plant, hairy roots, cell suspension etc.), expression type (stable, transient, transplastomic, tissue-specific, inducible) and product targeting (post-translational modifications and accumulation targets) [4]. Today, there are thousands of different recombinant proteins produced in plant systems in which we can also include antibody, vaccine, hormone or enzyme type pharmaceuticals. There is also an increasing number of companies producing commercial plant-based therapeutics (**Table 3**). Along with the obvious advantage of pharming in scale-up production, this approach also has a downside on downstream processing (including maintaining product quality, extraction and purification) due to the wide range of plant metabolites.

3.1 Monoclonal antibodies, viral antigens and vaccines

Recombinant monoclonal antibody (mAbs) production in pharming applications has grown rapidly since the first reported IgG1 antibody in transgenic tobacco by

Company Name	Product			References
iBio, Inc	IBIO-201 Prophylaxis of SARS-CoV-2 Spike Protein Fused Lichenase Protein / *Nicotiana benthamiana*	ACE2-FC Prophylaxis of SARS-CoV-2 Human Angiotensin Converting Enzyme 2 (ACE2) Fused To A Human Immunoglobulin G Fc fragment / *Nicotiana benthamiana*	IBIO-400 Prophylaxis of Classical Swine Fever (CSF) CSFV E2 Glycoprotein / *Nicotiana benthamiana*	[32]
Medicago Inc.	MT-7529 Prophylaxis of H7N9 Influenza / *Nicotiana benthamiana*	MT-2355 Prophylaxis Of Pertussis, Diphtheria, Tetanus, Poliomyelitis and Prophylaxis of Hib Infection in Infants / *Nicotiana benthamiana*	MT-2271 Prophylaxis of Seasonal Influenza / *Nicotiana benthamiana*	[33]
	MT-5625 Prophylaxis of Rotavirus Gastroenteritis / *Nicotiana benthamiana*	MT-2766 Prophylaxis of SARS-CoV-2 / *Nicotiana benthamiana*	MT-8972 Prophylaxis of H5N1 Influenza / *Nicotiana benthamiana*	
Icon Genetics	ZMapp Prophylaxis of Ebola virüs / *Nicotiana benthamiana*	Denka Prophylaxis of Norovirus / *Nicotiana benthamiana*	-	[34]
SemBioSys Genetics Inc.	Milano Production of apolipoprotein AI / *Carthamus tinctorius*	SBS-1000 Plant-Produced Insülin / *Carthamus tinctorius*	-	
Protailx	Elelyso Prophylaxis of type 1 Gaucher's disease / *Daucus carota*	PRX 102 Prophylaxis of Fabry disease / *Daucus carota*	PRX-105 Use in treatment of Organophosphorus poisoning / *Daucus carota*	[35]
Wuhan Healthgen Biotechnology Corp.	OsrhLF Recombinant Human Lactoferrin / *Oryza Sativa*	OsrHSA Human Serum Albumin / *Oryza Sativa*	OsrhEGF Human Epidermal Growth Factor / *Oryza Sativa*	[36]
	OsrhbFGF Human Basic Fibroblast Growth Factor / *Oryza Sativa*	rhIGF-1 LR3 Human Insulin-like Growth Factor-1 LR3 / *Oryza Sativa*	OsrhVEGF Human Vascular Endothelial Growth Factor / *Oryza Sativa*	
	OsrhKGF Keratinocyte Growth Factor / *Oryza Sativa*	OsrhLF Human Lactoferrin / *Oryza Sativa*	OsrhFN Human Fibronectin / *Oryza Sativa*	
	OsrhLYZ Human Lysozyme / *Oryza Sativa*	OsrhAAT Human α-1 Antitrypsin / *Oryza Sativa*	-	
Planet Biotechnology Inc.	PBI-220 Immunoadhesin of anthrax / *Nicotiana benthamiana*	(DPP4-Fc) Immunoadhesin of Middle East Respiratory Syndrome (MERS) / *Nicotiana benthamiana*	-	[37]

Company Name	Product			References
Merck KGaA	L1294 Lactoferrin / *Oryza Sativa*	A9731 Albumin / *Oryza Sativa*	L9545 Leukemia Inhibitory Factor / *Oryza Sativa*	[38]
	T3705 Transferrin / *Oryza Sativa*	L1667 Lysozyme / *Oryza Sativa*	616371 Aprotinin / *Nicotiana tabacum*	
	B0939 B Lymphocyte Activating Factor / *Nicotiana tabacum*	T3449 TrypZean® / *Zea Mays*	B0814 Bone Morphogenetic Protein 7 (BMP-7) / *Nicotiana tabacum*	
Angio-Proteomie	rAP-0487 Interleukin-12 p40 / *Nicotiana benthamiana*	rAP-2263 Growth Hormone / *Nicotiana benthamiana*	rAP-2375 Myostatin / *Nicotiana benthamiana*	[39]
Abbexa Ltd	abx263080 Bone Morphogenetic protein-7	abx263465 Fibroblast Growth Factor	abx260381 B-Cell-Activating Factor	[40]
SoyMeds, Inc.	soy-mSEB Prophylaxis of Staphylococcal Enterotoxin B / *Glycine max*	-	-	[41]
G+FLAS Life Sciences	The RBD Prophylaxis of SARS-CoV-2 / *Nicotiana benthamiana*	-	-	[42]
Kentucky BioProcessing, Inc.	V-101 Prophylaxis of Seasonal Flu / *Nicotiana benthamiana*	V-201 Prophylaxis of SARS-CoV-2 / *Nicotiana benthamiana*	-	[43]
Ventria Bioscience	VEN100 Prophylaxis of Clostridium Difficile	VEN BETA Prophylaxis of Enterotoxigenic *E. coli* (ETEC) / On Sale	-	[44]
Thermo Fisher Scientific Inc.	A35934 Leukemia Inhibitory Factor / *Hordeum vulgare*	-	-	[45]
ORF Genetics	ISOkine Production of Human Growth Factors & Cytokines / *Hordeum vulgare*	MESOkine Production of Anima-Like Growth Factors & Cytokines / *Hordeum vulgare*	DERMOkine EGF (Epidermal Growth Factor) / *Hordeum vulgare*	[46]

Table 3.
Plant-based therapeutic producing companies and commercial products.

Andrew Hiatt [30]. In the beginning, it was not an easy task due to the complicated nature of antibodies on basis of post-translational modifications, folding patterns and structural assembly. IgG is the simplest immunoglobulin structure, therefore only two plant genes are required to produce fully functional IgG in plant systems. On the other hand, IgA which has four heavy chains and four light chains requires the expression of four genes at the same time. Early antibody studies also considered producing antibody fragments, mini bodies, large single chains, single-chain variable fragments (scFvs), bispecific scFvs, diabodies and fusion proteins as well [47]. Some achievements on targeting the mAbs into different plant cell parts for accumulation followed in a short time. The first plant-derived IgM was produced in *N. tabacum* and targeted into chloroplasts. Constructed chimeric genes and the barley aleurone α-amylase signal peptide coding sequence which had already been used successfully to transport bacteriophage T4 lysozyme from transgenic tobacco cells to the intercellular spaces was introduced to *N. tabacum* to initiate the secretory pathway of chimeric IgM in transgenic plants. Subcellular localization of IgM, presented the assembly of the antibody in the endoplasmic reticulum and the targeted accumulation in chloroplasts. Assembly and targeting of complex foreign protein in the transgenic plant were shown through fusing the individual chains to a plant signal peptide [48]. Secretory IgA was also expressed recombinantly. Four transgenic *N. tabacum* plants were generated that expressed a murine monoclonal antibody kappa chain, a hybrid immunoglobulin A-G heavy chain, a murine joining chain and a rabbit secretory component, respectively. Sexual crosses were achieved among the transgenic tobaccos resulting expression of all four protein chains simultaneously. These chains were assembled into a functional, high molecular weight secretory IgA which recognized the native streptococcal antigen I/II cell surface adhesion molecule. Transgenic plants were suggested as suitable systems for large-scale production of assembled recombinant secretory IgA for passive mucosal immunotherapy since plants require a single cell to assemble secretory antibodies while mammalian cells require two different cell types [49]. Transgenic plant-derived antibodies (plantibodies) are thought to be particularly effective in topical immunotherapies which are based on the antigenic competition by using immunomodulators to induce hypersensitivity. In recent days, mABs are designed for various purposes as chemotherapeutics for cancer, antibody-mediated passive immunization against highly contagious infectious diseases as SARS and COVID-19, curing or slowing down disease progression, active immunization through antigens (**Table 4**).

mABs may aim cancer cells in different mechanisms. They can directly bind and flag cancer cells for immune cells prevent angiogenic properties, stimulate disruption of the cell membrane, block immune system inhibitors, retard cancer growth, act as chemotherapy or radiotherapy agent carrier. Targeted antigens are generally related to growth and differentiation including epidermal growth factor receptor (EGFR), carcinoembryonic antigen (CEA), human epidermal growth factor receptor 2 (HER2), vascular endothelial growth factor (VEGF), VEGF receptor (VEGFR), fibroblast activation protein (FAP). mAbs are not native products of plant tissues. However, the transgenic plant approach provides the capacity to induce production and modification of mAbs through posttranslational modifications as glycosylation to enhance therapeutic efficacy. Pembrolizumab which is an anti-human PD-1 monoclonal mAb has been produced in wild-type *Nicotiana benthamiana* through transient expression. 344.12 ± 98.23 μg/g fresh leaf weight Pembrolizumab accumulation was obtained after 4 days of agroinfiltration. Molecular characterization of plant-based Pembrolizumab was compared to mammalian cell-based commercial counterpart Keytruda®. Physicochemical properties of plant-based Pembrolizumab were found comparable to Keytruda® with similar secondary and tertiary structures. Both products presented no aggregation differences and binding efficacy to PD-1 protein and inhibitory activity between programmed cell death 1 (PD-1) and programmed cell death ligand 1 (PD-L1) interaction. In this respect, researchers concluded that plant-produced Pembrolizumab could induce IL-2 and IFN-γ production and plant-based production of functional Pembrolizumab can be utilized for immunotherapy purposes [50]. In another anticancer approach, heavy and light chains of mAb BR55-2 were expressed separately and assembled in plant cells of transgenic tobacco plants (*N. tabacum* cv. LAMD609). Production was as high as 30 mg/kg of fresh leaves in the first generation of plants. Like in mammalian counterpart, the Fc

Plant	Target	Protein	Expression system	Transformation method	Efficiency	Purification	References
Nicotiana benthamiana	Cancer	Pembrolizumab (PD-1) monoclonal antibody	Transient Expression	CaMV 35S promoter/ *Agrobacterium tumefaciens*-mediated	344.12 ± 98.23 μg/g FLW	Size Exclusion Chromatographyband protein A Affinity Chromatography, The ÄKTA Pure Fast Protein Liquid Chromatography (FPLC)	[50]
Glycoengineered (ΔXFT) Nicotiana benthamiana	Chikungunya virus (CHIKV)	The Monoclonal Antibody anti-CHIKV	MagnICON-based Transient Expression	*Agrobacterium tumefaciens*-mediated	130 μg/g FLW	Low pH Precipitation and Protein A Chromatography	[51]
Nicotiana benthamiana	Severe acute respiratory syndrome coronavirus-2 (SARS-CoV-2)	The Monoclonal Antibody CR3022	Transient Expression	CaMV 35S promoter/ *Agrobacterium tumefaciens*-mediated	130 μg/g FLW	Ni affinity Chromatography	[52]
Nicotiana benthamiana	Zika virus	The Monoclonal Antibody 2A10G6	BeYDV-Based Transient Expression	CaMV 35S promoter / -	3–5 g/kg FLW	G column Chromatography	[53]
Nicotiana benthamiana	Hand-Foot-Mouth Disease (HFMD)	The Monoclonal Antibody D5	Transient Expression	CaMV 35S promoter/ *Agrobacterium tumefaciens*-mediated	50 μg/g FLW	0.45 μm Membrane Filter and protein A Chromatography	[54]
Oryza sativa	Porphyromonas gingivalis	MAb specific for FimA (Fimbrial protein fimbrillin)	Nucleer / Stable Expression	Rice Amylase 3D (Ramy3D) promoter / Particle Bombardment	3.44 μg/g FLW	G-affinity Chromatography	[55]
Nicotiana benthamiana	Dengue Virus (DENV)	The Monoclonal Antibody E60	Transient Expression	magnICON® / *Agrobacterium tumefaciens* mediated	0.8–4.8 mg/g FLW	Ammonium Sulfate precipitation and Protein A Affinity Chromatography.	[56]
Nicotiana benthamiana	AIDS (acquired immunodeficiency syndrome)	The Monoclonal Antibody 2G12	Nucleer / Stable Expression	CaMV 35S promoter/ *Agrobacterium tumefaciens*-mediated	7-10 μl/ml Leaf Extract	PDF4 Filter Purification	[57]
Nicotiana tabacum	AIDS (acquired immunodeficiency syndrome)	The Monoclonal Antibody 2G12	Transient Expression	CaMV 35S promoter/ *Agrobacterium tumefaciens*-mediated	-	Protein A Chromatography	[58]

DOI: http://dx.doi.org/10.5772/intechopen.102004

Plant	Target	Protein	Expression system	Transformation method	Efficiency	Purification	References
Zea mays	AIDS (acquired immunodeficiency syndrome)	The Monoclonal Antibody 2F5	Transient Expression	CaMV 35S promoter/ Particle Bombardment	0.61 ± 0.28 µg/ ml Seed extract	-	[59]
Nicotiana tabacum	AIDS (acquired immunodeficiency syndrome)	The Monoclonal Antibody 4E10	Nucleer / Stable Expression	CaMV 35S promoter / *Agrobacterium rhizogenes*-mediated	10.43 RDW	-	[60]
Nicotiana benthamiana	West Nile virus (WNV)	The Monoclonal Antibody pHu-E16	(TMV)-based Transient Expression	*Agrobacterium tumefaciens*-mediated	0.8 mg/g FLW	Ammonium Sulfate Precipitation and Protein A Chromatography	[61]
Nicotiana tabacum	Hepatitis B Virus	Anti-HBsAg Monoclonal Antibody	Nucleer / Stable Expression	Phaseolin promoter	6.5 mg/g of Seed	Protein A Affinity Chromatography, Size-exclusion chromatography (SEC) and High-performance liquid chromatography (HPLC)	[62]
Nicotiana tabacum	Rabies Virus (RABV)	The Monoclonal Antibody E559	Nucleer / Stable Expression	-	1.8 mg/kg FLW	Protein A Chromatography	[63]
Lactuca sativa	Ebola and West Nile viruses	The Monoclonal Antibodies 6D8 and hE16	MagnICON-based Transient Expression	*Agrobacterium tumefaciens*-mediated	0.27 mg/g	DEAE anion-exchange chromatography with DEAE Sepharose FF 26/20 resin	[64]
Nicotiana tabacum	Streptococcus-mediated dental caries	The Monoclonal Antibody Guy's 13	Nucleer / Stable Expression	CaMV 35S promoter / *Agrobacterium rhizogenes*-mediated	58 µg/g RDW	Affinity Chromatography	[60]
Nicotiana tabacum	Cancer	The Monoclonal Antibody BR55-2	Nucleer / Stable Expression	CaMV 35S promoter/ *Agrobacterium tumefaciens*-mediated	30 mg/kg FLW	Protein A Chromatography	[65]

Table 4.
Monoclonal antibodies produced from transgenic plants.

domain of the plant mAbP presented a similar binding to FcγRI receptor (CD64) and bound to both SK-BR3 breast cancer cells and SW948 colorectal cancer cells, specifically. This plant-derived BR55-2 also inhibited SW948 tumor growth in nude mice, efficiently and was suggested as a possible immunotherapy option [65]. Both samples represent IgG-type plant-derived antibodies. However, more complex IgA-type plant-derived products are also presented with better efficacy, recently. The secretory component (SC) of immunoglobulin A (SIgA), which is an efficient therapeutic antibody against mucosal pathogens, was successfully expressed in *A. thaliana*. The expression level of SC was increased in the plant system through the insertion of endoplasmic reticulum retention signal peptide, KDEL (Lys-Asp-Glu-Leu), into a binary vector with translational enhancer and an efficient terminator [66]. This approach was also reported as useful against food poisoning causing *E. coli* virulence factor Shiga toxin through recombinant IgA (S-hyIgA) produced in transgenic *A. thaliana* plants [67].

Plant-derived antibodies and viral antigens were also targeted for highly contagious infectious diseases as SARS, Ebola, Zica, Hepatitis B, AIDS and even the most recent COVID-19 (**Tables 5** and **6**). SARS-CoV-2 is a single-stranded RNA-enveloped virus, which has 29,881 bp genome encoding 9860 amino acids belonging to structural proteins, spike (S), envelope (E), membrane (M) and nucleocapsid (N) and 16 non-structural proteins (such as 3-chymotrypsin-like protease, papain-like protease and RNA-dependent RNA polymerase) and 9 accessory proteins. N protein which is the most abundant viral protein shown to be highly conserved in CoV families. It is highly immunogenic during CoV infections. It is a major target for antibody responses and also contains T cell epitopes [82, 83]. Post-transcriptional gene silencing suppressor p19 protein from tomato bushy stunt virus substantially demonstrated the transient expression of recombinant SARS-CoV nucleocapsid (rN) protein in *Nicotiana benthamiana*. The rN protein accumulated up to a concentration of 79 μg per g fresh leaf weight in the agrobacteria-infiltrated plant leaf after the third day of infiltration. BALB/c mice were intraperitoneally vaccinated with pre-treated plant extract emulsified in Freund's adjuvant and plant-expressed recombinant SARS-CoV N protein-induced strong humoral and cellular responses in mice [74]. Also, iBio company developed a plant-derived vaccine targeting the N protein in their IBIO-202 program which is under pre-clinical trials. M and E proteins contribute very low on protection owing to their small ectodomains for immune cell recognition and small molecular sizes and poorly immunogenic for humoral responses. Glycosylated S proteins cover the SARS-CoV-2 surface and bind to the host cell receptor angiotensin-converting enzyme 2 (ACE2) during viral cell entry. Hence, S protein is the main viable vaccine target against the ongoing pandemic for the time being [84]. Rapid production of SARS-CoV-2 receptor-binding domain (RBD) and spike-specific monoclonal antibody CR3022 were achieved in *Nicotiana benthamiana*. Both RBD and mAb CR3022 were transiently produced with the highest expression level of 8 μg/g and 130 μg/g leaf fresh weight respectively at 3 days post-infiltration. The plant-produced RBD exhibited specific binding to the SARS-CoV-2 receptor, angiotensin-converting enzyme 2 (ACE2) [52]. In July 2020, phase 1 study was initiated for a plant-derived virus-like particle (VLP) vaccine candidate called CoVLP which expresses the SARS-CoV-2 spike glycoprotein (CoVLP: NCT04450004). It has been evaluated through 21 days apart from two doses of 3.75 μg, 7.5 μg or 15 μg vaccine alone or with AS03 or CpG1018 adjuvants in healthy adults 18–55 years of age. Based on the available data two-dose schedule of CoVLP at 3.75 μg per dose adjuvanted with AS03 has been carried forward into ongoing phase 2/3 studies in Canada and the United States, with planned expansion to additional countries in Latin America and Europe [85]. In December 2020, the Kentucky BioProcessing company announced its phase 1 plant-derived

Plant	Disease	Antigent	Expression system	Transformation method	Efficiency	immunization	İmmun responce	References
Nicotiana Tabacum / Lactuca sativa	type 1 diabetes	human proinsulin (hpINS) fusion protein with CTB	Chloroplast / Stable Expression	psbA promoter / biolistics	3.33–15.3 mg/g CTB-hpINS (12.9–24.4 of TSP)	Oral / 250-500 μg	Decrease IL-10	[68]
Spinacia oleracea /	Type-2 diabetes (T1D)	glutamic acid decarboxylase (GAD65)	Nucleer / magnICON® Based Transient Expression	BAK/ *Agrobacterium tumefaciens* mediated	SO: 544 μg/g FW	-	-	[69]
Beta vulgaris	Type-2 diabetes (T1D)	glutamic acid decarboxylase (GAD65)	Nucleer / magnICON® Based Transient Expression	BAK/ *Agrobacterium tumefaciens* mediated	BV: 113 μg/g FW	-	-	[69]
Nicotiana tabacum	Parkinson's disease (PD)	LTB-Syn chimeric protein	Nucleer/ Stable Expression	CaMV 35S promoter/ *Agrobacterium tumefaciens*-mediated	0.27 μg/g	İntraperitoneal / 10 μg	Induced IgG	[70]
Oryza sativa	Allergic Asthma	*Dermatophagoides pteronyssinus* allergen (Der p 1)	Nucleer/ Stable Expression	GluB-1 promoter / *Agrobacterium tumefaciens* mediated	7.5% TSP	Oral / 0.5 or 5 mg, intraperitoneal / 6.8 mg	Decrease IgE IgG2a and IgG2b, inhibition of IL-4, IL-5, and IL-13	[71]
Lycopersicon esculentum	Alzheimer's disease,	Human β-amyloid Aβ	Nucleer/ Stable Expression	CaMV 35S promoter/ *Agrobacterium tumefaciens*-mediated	0.7% TSP	Oral / 5 g	NA	[72]
Oryza sativa	Pollen Allergy	The T cell epitope peptides of Cry j I and Cry j II	Nucleer/ Stable Expression	GluB-1 promoter / *Agrobacterium tumefaciens* mediated	0.5% TSP	Oral / 70 μg	Decrease IgE IgG2a and IgG2b, inhibition of IL-4, IL-5, and IL-13	[73]

Table 5.
Plant-based vaccines for non-infectious diseases.

Plant	Pathogen/Disease	Antigent	Expression system	Efficiency	Adjuvant	Immunization	Immun responce	References
Nicotiana benthamiana	SARS-CoV	SARS-CoV nucleocapsid (rN) protein	Nucleer / Transient Expression	0.8–1% of the TSP	With Complete Freund's Adjuvant/ Incomplete Freund's Adjuvant	İntraperitoneal / 500mg fresh leaves	Induced IgG1 and IgG2/ Increase IFN-_ and IL-10/ not chanced IL-2 and IL-4	[74]
Nicotiana tabacum	Ebola virus (EBOV)	Envelope-Associated Protein VP40	Nucleer / Stable Expression	2.6 μg/g FW	With Complete Freund's Adjuvant / Incomplete Freund's Adjuvant	Ebible (25 ng) / Subcutaneous (125 ng)	Induced IgM, IgG and intestinal IgA	[75]
Nicotiana benthamiana	Zika virus (ZIKV)	Envelope (E) Protein	Nucleer / Transient Expression	160 μg/g FW	With Aluminium Hydroxide Gel Adjjuvant	Subcutaneous / 50 μg x 24	Induced IgG1 and IgG2, Increase IFN-γ, IL-4 and IL-6	[76]
Oryza sativa	Vibrio cholerae	Cholera Toxin B-Subunit	Nucleer / Stable Expression	NA	Witout Adjuvant	Edible / 150 mg seed	induced IgG and mucosal IgA	[77]
Nicotiana benthamiana	Influenza A H1N1	Soluble Protein H1/ H1-VLP	Nucleer / Transient Expression	NA	-	-	Induced CD4+ and CD8+ T cells	[78]
Nicotiana benthamiana	Influenza A H5N1	The matrix protein 2 ectodomain (M2e) fused to N-terminal proline-rich domain (Zera®) of the γ-zein protein of maize	Nucleer / Transient Expression	125–205 mg/ kg FW	Without Adjuvant	İntramuscular / 4.5 μg	Induced IgG	[79]
Nicotiana tabacum / Lycopersicon esculentum	Yersinia pestis / Plague	The major capsular protein F1-V antigen fused	Nucleer NT:Transient Expression / LE:Stable Expression	NT: 1–4% FW LE: 4–10% mg DW	With Adjuvant NT: Aluminum Hydroxide T / LE: Cholera Toxin	NT: Subcutaneous (10μg purified) / LE: Edible (2 g fruit)	Induced Serum IgG1, IgG2a and mucosal IgA	[80]

DOI: http://dx.doi.org/10.5772/intechopen.102004

Plant	Pathogen/Disease	Antigent	Expression system	Efficiency	Adjuvant	Immunization	Immun responce	References
Nicotiana benthamiana	Flavivirus / Yellow fever (YF)	YF virus envelope protein (YFE) fusion to the bacterial enzyme lichenase (YFE-LicKM)	Nucleer / Transient Expression	NA	With Alhydrogel Adjuvant	İntramuscular 5 μg x3/ 5 μg x 2 / 30 μg x 3	Induced IgG İncreased IFNγ	[81]

FW: Fresh Weight, TSP: Total Soluble Protein, NA: Not Available, DW: Dry Weight. NT: Nicotiana tabacum, LE: Lycopersicon esculentum. CaMV: Cauliflower Mosaic Virus

Table 6.
Plant-based vaccines for pandemics and epidemics infectious diseases.

cVLP vaccine. CPG adjuvant vaccine was administered in healthy adult subjects in two age groups, Part A (18–49 years) and Part B (50–85 years). The company also produces seasonal influenza vaccines through *Nicotiana benthamiana* plants [86]. BaiyaPharming™ produced a subunit-based plant-derived vaccine against SARS-CoV-2 in *N. benthamiana*. Baiya SARS-CoV-2 Vax 1 was chosen between six candidates which showed better immunogenicity in mice and monkeys. Baiya Phytopharm expects to initiate human trials by September 2021 [87]. Plant-derived vaccines for humans and animals stand out as a viable alternative that can be used to overcome the barriers of conventional vaccines. Within the scope of transgenic plants, it is possible to produce cost-effective, immunogenic and safer vaccines in plants with an enhanced amount, effective isolation and purification methods. Correct use of adjuvants along with the production of recombinant vaccine antigens also seems equally crucial for the future of this technology.

3.2 Replacement human proteins

Plants are usually referred to as molecular factories to provide humans with many useful molecules for many purposes. In the last decades, it has also become available to produce specific heterologous proteins as a replacement in humans. In this manner, the first plant-derived pharmaceutical was human growth hormone, which was expressed in transgenic tobacco as a fusion with *Agrobacterium* nopaline synthase enzyme in 1986 [88]. Today, recombinant human proteins are a considerable part of FDA-approved biotechnological drugs and recombinant plant-derived proteins are extended in many categories as industrial enzymes, research intended technical proteins, nutritional supplements and polymers as well as antibodies and vaccines which were mentioned in the previous section. Replacement human proteins include products as growth hormone, HSA, α-interferon, erythropoietin (EPO), human secreted alkaline phosphatase, aprotinin, collagen, α1-antitrypsin and more (**Table 7**).

Human growth hormone (hGH) has various biological functions on protein synthesis, cell proliferation and metabolism. After the first successful plant-derived production, many different strategies were achieved. In a recent approach, a synthetic hGH gene (shGH) has been synthesized in a plant expression vector under the control of the rice amylase 3D (Ramy3D) promoter. The plant expression vector was introduced into rice calli (*O. sativa* L.) via the particle bombardment transformation method. The shGH protein expression was verified and quantified as 57 mg/L in the transgenic rice cell suspension medium. Biological activities of the shGH were found similar to the conventional *E. coli*-derived recombinant hGH. Likewise, many different plant tissues and expression systems are suggested as effective hGH production replacements [101].

Human serum albumin (HSA) is the most abundant protein in human blood plasma. HSA is a soluble, globular, unglycosylated, monomeric multidomain protein. The single polypeptide of HSA consists of 585 amino acids with a range of structural configurations that fold into three helical domains [104]. It is also the first full-size native human protein expressed in plants and there is more than 500 tons annual demand. The transient expression level of the HSA gene in different genotypes was achieved in many plants. Recently, the A. tumefaciens strains LB4404 and GV3101 containing pBI121-HSA binary vector were infiltered in *Nicotiana benthamiana* and *N. tabacum* varieties. The bioactive HAS expression in tobacco leaves through the expression of the HSA gene in the plant system is suggested as the first transient expression success in literature [105].

α-interferon is the first human pharmaceutical protein produced in rice. The plasmid pIG3031 containing human α-interferon cDNA and the neomycin

DOI: http://dx.doi.org/10.5772/intechopen.102004

Plant	Disease	Protein	Expression system	Transformation method	Efficiency	Purification	References
Nicotiana benthamiana	Osteogenic differentiation	Human Dentin Matrix Protein 1 (hDMP1)	Transient Expression	CaMV 35S promoter/ *Agrobacterium tumefaciens*-mediated	0.3 μg/g FW	Ni affinity chromatography	[89]
Nicotiana benthamiana	In The Diagnosis Of Rift Valley Fever Virus (RVFV)	Nucleocapsid Protein (N-protein)	Nucleer/ Stable Expression	CaMV 35S promoter/ *Agrobacterium tumefaciens*-mediated	500–558 mg/kg FLW	Ammonium Sulphate Precipitation, Nickel Affinity Column Chromatography and 6xhis-Tag Affinity Chromatography	[90]
Nicotiana tabacum	Anemia	Erythropoietin (rhEPO)	Nucleer/ Stable Expression	CaMV 35S promoter/ *Agrobacterium rhizogenes*-mediated	66.75 pg/ mg medium TSP	Sephadex 25 columns, Ni-TED columns, HPLC	[91]
Nicotiana benthamiana	Dental Bone Regeneration	Osteopontin (hOPN)	Transient Expression	CaMV 35S promoter / *Agrobacterium tumefaciens*-mediated	100 ng/g FW	Ni affinity chromatography	[92]
Physcomitrella patens	Fabry disease	α-galactosidase A	Nucleer / Stable Expression	PEG-based	0.5 mg/ml TSP	Butyl-650 M, DEAE, S chromatography	[93]
Nicotiana benthamiana	Gaucher's disease (GCase)	Human Glucocerebrosidase	Nucleer/ Stable Expression	CaMV 35S promoter/ *Agrobacterium tumefaciens*-mediated	68 μg/g FW	Concanavalin A (Con A) Agarose Column and Hydrophobic Interaction Chromatography (Phenyl-650C)	[94]
Nicotiana tabacum	Pompe disease	Acid Alpha Glucosidase (GAA)	Chloroplasts / Stable Expression	psbA promoter / Biolistic Mediated	190 μg/g of DW	-	[95]
Salvia miltiorrhiza	Angiogenesis, and Tissue Repair	Human acidic fibroblast growth factor 1 (FGF-1)	Nucleer/ Stable Expression	CaMV 35S promoter/ *Agrobacterium tumefaciens*-mediated	272 ng/g FW	-	[96]

Plant	Disease	Protein	Expression system	Transformation method	Efficiency	Purification	References
Helianthus annuus	Cardiovascular and Cerebrovascular Thrombus Diseases	Lumbrokinase (LK)	Nucleer/ Stable Expression	napA promoter / *Agrobacterium tumefaciens*-mediated	5.1 g/kg of Seed	Nothing	[97]
Nicotiana tabacum	Drug Carrier	Hydrophobin	Nucleer/ Stable Expression	CaMV 35S promoter/ *Agrobacterium tumefaciens*-mediated	16.5% of TSP	Surfactant-Based Aqueous Two-Phase Separation	[98]
Nicotiana benthamiana	Anthrax	Anthrax Receptor Decoy Protein (immunoadhesin), CMG2-Fc	Transient Expression	CaMV 35S promoter / *Agrobacterium tumefaciens*-mediated	1.5% of TSP	-	[99]
Nicotiana benthamiana	Cardiopulmonary Bypass	r-aprotinin	GENEWARE® mediated Transient Expression	CP promoter / *Agrobacterium tumefaciens*-mediated	25 mg/kg FW	Size-exclusion chromatography (SEC) and High-performance liquid chromatography (HPLC)	[100]
Oryza sativa	Dwarfism, bone fractures, skin burns, and bleeding ulcers	Human growth hormone (hGH)	Nucleer/ Stable Expression	Rice Amylase 3D (Ramy3D) promoter / Particle Bombardment Transformation	57 mg/l medium TSP	-	[101]
Glycine max	Cardiovascular and Neurodegenerative Diseases	Fibroblast Growth Factor (Bfgf)	Nucleer/ Stable Expression	G1 (glycinin) promoter / *Agrobacterium tumefaciens*-mediated	2.3% of TSP	heparin-Sepharose CL-6B affinity chromatography	[102]
Nicotiana tabacum	Marker Used In Diagnosis	Placental Alkaline Phosphatase (SEAP)	Nucleer/ Stable Expression	CaMV 35S promoter/ *Agrobacterium tumefaciens*-mediated	3% of TSP	-	[103]

Table 7.
Replacement human proteins produced from transgenic plants.

phosphotransferase II coding sequence was introduced to Indica rice protoplasts via lipofection-mediated transformation. Transgenic plants were regenerated from transformed calli. Extracts of transgenic cell cultures and plants presented apparent interferon activity proven by the resistance of human amniotic cell lines to viral infection in the presence of plant extracts. This production encouraged many other strategies as human α-interferon cDNA was correctly expressed in rice cells [106].

Erythropoietin (EPO) is the first human replacement protein produced in tobacco suspension cells. EPO is a cytokine that regulates and maintains the physiological level of circulating erythrocytes. The survival of erythroid precursor cells is also achieved through EPO. It stimulates the proliferation and differentiation of the precursor cells by plasma membrane EPO-receptor interactions. The first transgenic approach introduced human Epo cDNA via *A. tumefaciens*-mediated gene transfer to tobacco BY2 cells (*N. tabacum* L. cv. Bright Yellow 2). EPO is a heavily glycosylated protein, therefore glycosylation of tobacco-derived EPO by smaller oligosaccharides led the molecule to remain attached to the cell wall. However, it induced the differentiation and proliferation of erythroid cells in in vitro biological activity trials [107]. Recently, mammalian cell-derived recombinant human erythropoietin (rhuEPOM) is a multimodal neuroprotectant in experimental stroke models. However, the rhuEPOM clinical trials were terminated due to the increased risk of thrombosis, largely ascribed to its erythropoietic function. A rhuEPO derivative without sialic acid residues was produced in a plant-based expression system which is lacking sialylation capacity to produce asialo-rhuEPO^P. Repeated intravenous injection (44 μg/kg bw) in mice presented no increase in hemoglobin levels and red blood cells. Hence, Asialo-rhuEPO^P that lacks erythropoietic activity and immunogenicity suggested as a great multimodal neuroprotectant for stroke treatment [108].

Pathogen contamination risk of animal-derived collagen initiated the need for safe recombinant production of this complex molecule. Collagen is the first human structural-protein polymer produced in a plant-derived system. The use of the tobacco plant as a novel expression system for the production of human homotrimeric collagen I was achieved in 2000. cDNA encoding the human proalpha1(I) chain was introduced to tobacco. Expressed recombinant procollagen has been folded to stable homotrimeric triple helix-shaped collagen as in animal cells [109]. *A. tumefaciens*-mediated transient expression of the recombinant hydroxylated homotrimeric collagen in tobacco plants that are co-transformed with a human type I collagen and a chimeric proline-4-hydroxylase (P4H) improved the quality of collagen by enhancing thermostability to 37°C [110].

Human 1-antitrypsin (AAT) is a 394-amino-acid glycoprotein that inhibits the activity of the serine protease neutrophil elastase. Healthy individuals control elastase activity by producing sufficient quantities of AAT into the bloodstream. When the circulating concentration drops below 15%, various diseases as emphysema, hepatitis and skin disorders occur. Various expression systems, cell types and tissues of plants were used for molecular pharming earlier. However, rice suspension cells were used for the molecular farming of recombinant human 1-antitrypsin (rAAT) in biologically active form for the first time in 1999. Transformation of rice callus tissues with a p3D-AAT expression vector containing the cDNA for mature human AAT protein was achieved. The promoter, signal peptide and terminator of a rice-amylase gene Amy3D, which tightly controls simple sugars such as sucrose, regulated expression and secretion of rAAT. Expression of the rAAT was initiated by removing sucrose from the cultured media or by allowing the rice suspension cells to deplete sucrose catabolically. For that time being, the rice cell culture system clearly contributed to the molecular pharming field.

As a concluding summary, plants have various upsides against traditional microbial and animal cell culture systems in respect to molecular pharming. Some of these include cost efficiency, easier up-scale production, absence of human pathogens and accurate maturation (folding and assembly) of proteins. The potential of molecular pharming in plants attracts more and more entrepreneurs following the numerous successful products and companies. Achieving secretory properties and targeting of plant products also surpass many other production systems. The use of plants as bioreactors is well known, and its applications are increasing for both recombinant protein expression and recombinant pharmaceutical production. This chapter showcases the various plant biotechnology application and strategies as applied in the production of recombinant antibodies, antigens, therapeutic proteins and enzymes, that are used in the treatment of various diseases. Now with the COVID-19 pandemic, more than ever this approach is taking center stage. This is so important as it will reveal the hidden treasure that transgenic plants already offered but mired in the genetically modified organism debate and therefore rejected before these other applications beyond food could be readily realized. Indeed, different systems exist within the system utilizing many different parts and tissues of the plant to produce products. Furthermore, this system is more advantageous in a eukaryotic system that performs post-translational modification as would animal and yeast cells thus yielding the final desired therapeutic product comparable to what is already produced by humans, for example.

Author details

Alp Ayan*, Sinan Meriç, Tamer Gümüş and Çimen Atak
Faculty of Science and Letters, Department of Molecular Biology and Genetic, Istanbul Kultur University, Istanbul, Turkey

*Address all correspondence to: a.ayan@iku.edu.tr

References

[1] Meriç S, Ayan A, Atak Ç. Molecular abiotic stress tolerans strategies: From genetic engineering to genome editing era. Abiotic Stress Plants. 2020:91-126. DOI: 10.5772/INTECHOPEN.94505

[2] Çakır Ö, Meriç S, Arı Ş. GMO Analysis Methods for Food: From Today to Tomorrow. In: Spizzirri GU, Cirillo G, editors. Food Safety: Innovative Analytical Tools for Safety Assessment. New York: WILEY-Scrivener Publishing; 2017. pp. 123-179

[3] ISAAA. GM Approval Database | GMO Database | GM Crop Approvals. (n.d.). Available from: ISAAA.org

[4] Meriç S, Gümüş T, Ayan A. Plant-based Vaccines: The Future of Preventive Healthcare?, Botany - Recent Advances and Applications, Bimal Kumar Ghimire: IntechOpen; June 8th 2021. DOI: 10.5772/intechopen.97861. Available from: https://www.intechopen.com/chapters/76930

[5] Webster GR, Teh AY-H, Ma JK-C. Synthetic gene design—The rationale for codon optimization and implications for molecular pharming in plants. Biotechnology and Bioengineering. 2017;**114**:492-502. DOI: 10.1002/BIT.26183

[6] Tschofen M, Knopp D, Hood E, Stöger E. Plant molecular farming: Much more than medicines. Annual Review of Analytical Chemistry. 2016;**9**:271-294. DOI: 10.1146/ANNUREV-ANCHEM-071015-041706

[7] Lee J-H, Park D-Y, Lee K-J, Kim Y-K, So Y-K, Ryu J-S, et al. Intracellular Reprogramming of Expression, Glycosylation, and Function of a Plant-Derived Antiviral Therapeutic Monoclonal Antibody. PLoS One. 2013;**8**:e68772. DOI: 10.1371/JOURNAL.PONE.0068772

[8] Gomord V, Fitchette A-C, Menu-Bouaouiche L, Saint-Jore-Dupas C, Plasson C, Michaud D, et al. Plant-specific glycosylation patterns in the context of therapeutic protein production. Plant Biotechnology Journal. 2010;**8**:564-587. DOI: 10.1111/J.1467-7652.2009.00497.X

[9] Stoger E, Fischer R, Moloney M, Ma JK-C. Plant molecular pharming for the treatment of chronic and infectious diseases. Annual Review of Plant Biology. 2014;**65**:743-768. DOI: 10.1146/Annurev-Arplant-050213-035850

[10] Dirisala VR, Nair RR, Srirama K, Reddy PN, Rao KRSS, Kumar NSS, et al. Recombinant pharmaceutical protein production in plants: unraveling the therapeutic potential of molecular pharming. Acta Physiologiae Plantarum. 2016;**39**:1-9. DOI: 10.1007/S11738-016-2315-3

[11] Dobhal S, Chaudhary VK, Singh A, Pandey D, Kumar A, Agrawal S. Expression of recombinant antibody (single chain antibody fragment) in transgenic plant Nicotiana tabacum cv. Xanthi. Molecular Biology Reports. 2013;**40**:7027-7037. DOI: 10.1007/s11033-013-2822-x

[12] Ting-Kuo Huang H, Michael P, Karen M. Semicontinuous bioreactor production of a recombinant human therapeutic protein using a chemically inducible viral amplicon expression system in transgenic plant cell suspension cultures. Biotechnology and Bioengineering. 2010;**106**:408-421. DOI: 10.1002/BIT.22713

[13] Fischer R, Schillberg S, Buyel JF, Twyman RM. Commercial aspects of pharmaceutical protein production in plants. Current Pharmaceutical Design. 2013;**19**:5471-5477. DOI: 10.2174/1381612811319310002

[14] Obembe OO, Popoola JO, Leelavathi S, Reddy SV. Advances in plant molecular farming. Biotechnology Advances. 2011;**29**:210-222. DOI: 10.1016/J.BIOTECHADV.2010.11.004

[15] Ahmad P, Ashraf M, Younis M, Hu X, Kumar A, Akram NA, et al. Role of transgenic plants in agriculture and biopharming, Biotechnology Advances. 2012;**30**(3):524-540. ISSN 0734-9750. DOI: 10.1016/j.biotechadv.2011.09.006

[16] Biotech Information Resources—ISAAA.org. (n.d.). Available from: https://www.isaaa.org/resources/default.asp [Accessed: October 22, 2021]

[17] Agarwal P, Agarwal PK, Joshi AJ, Sopory SK, Reddy MK. Overexpression of PgDREB2A transcription factor enhances abiotic stress tolerance and activates downstream stress-responsive genes. Molecular Biology Reports. 2009;**372**(37):1125-1135. DOI: 10.1007/S11033-009-9885-8

[18] Zhao X, Yang X, Pei S, He G, Wang X, Tang Q, et al. The Miscanthus NAC transcription factor MlNAC9 enhances abiotic stress tolerance in transgenic Arabidopsis. Gene. 2016;**586**:158-169. DOI: 10.1016/J.GENE.2016.04.028

[19] Gao SQ, Chen M, Xu ZS, Zhao CP, Li L, Xu HJ, et al. The soybean GmbZIP1 transcription factor enhances multiple abiotic stress tolerances in transgenic plants. Plant Molecular Biology. 2011;**75**:537-553. DOI: 10.1007/s11103-011-9738-4

[20] Djemal R, Khoudi H. TdSHN1, a WIN1/SHN1-type transcription factor, imparts multiple abiotic stress tolerance in transgenic tobacco. Environmental and Experimental Botany. 2016;**131**:89-100. DOI: 10.1016/J.ENVEXPBOT.2016.07.005

[21] Kathuria H, Giri J, Nataraja KN, Murata N, Udayakumar M, Tyagi AK. Glycinebetaine-induced water-stress tolerance in codA-expressing transgenic indica rice is associated with up-regulation of several stress responsive genes. Plant Biotechnology Journal. 2009;**7**:512-526. DOI: 10.1111/J.1467-7652.2009.00420.X

[22] Montero-Barrientos M, Hermosa R, Cardoza RE, Gutiérrez S, Nicolás C, Monte E. Transgenic expression of the Trichoderma harzianum hsp70 gene increases Arabidopsis resistance to heat and other abiotic stresses. Journal of Plant Physiology. 2010;**167**:659-665. DOI: 10.1016/J.JPLPH.2009.11.012

[23] Amara I, Capellades M, Ludevid MD, Pagès M, Goday A. Enhanced water stress tolerance of transgenic maize plants over-expressing LEA Rab28 gene. Journal of Plant Physiology. 2013;**170**:864-873. DOI: 10.1016/J.JPLPH.2013.01.004

[24] Li F, Ni H, Yan W, Xie Y, Liu X, Tan X, et al. Overexpression of an aquaporin protein from Aspergillus glaucus confers salt tolerance in transgenic soybean. Transgenic Research. 2021 Dec;**30**(6):727-737. DOI: 10.1007/s11248-021-00280-9. Epub 2021 Aug 30. PMID: 34460070

[25] Ferry N, Jouanin L, Ceci LR, Mulligan EA, Emami K, Gatehouse JA, et al. Impact of oilseed rape expressing the insecticidal serine protease inhibitor, mustard trypsin inhibitor-2 on the beneficial predator Pterostichus madidus. Molecular Ecology. 2005;**14**:337-349. DOI: 10.1111/J.1365-294X.2004.02381.X

[26] Zhu T et al. Ethylene-induced NbMYB4L is involved in resistance against tobacco mosaic virus in Nicotiana benthamiana. Hoboken, NJ: Wiley; 2021

[27] Wen L, Zhong J, Cui Y, Duan Z, Zhou F, Li C, et al. Coexpression of I. variabilis-EPSPS* and WBceGO-B3S1 genes contributes to high glyphosate

tolerance and low glyphosate residues in transgenic rice. Journal of Agricultural and Food Chemistry. 2021;**69**:7388-7398. DOI: 10.1021/ACS.JAFC.1C00880

[28] Mundt T, Albrecht L, Albrecht A, Krenchinski F. Growth and Development of Soybean Plants with the Pat Gene under different Glufosinate Rates. International Journal of Agriculture and Biology. 2021;**26**:217-223. DOI: 10.17957/IJAB/15.1827

[29] Stieger M, Neuhaus G, Momma T, Schell J, Kreuzaler F. Self assembly of immunoglobulins in the cytoplasm of the alga Acetabularia mediterranea. Plant Science. 1991;**73**:181-190. DOI: 10.1016/0168-9452(91)90027-6

[30] Hiatt A, Caffferkey R, Bowdish K. Production of antibodies in transgenic plants. Nature. 1989;**342**:76-78. DOI: 10.1038/342076a0

[31] Sijmons PC, Dekker BMM, Schrammeijer B, Verwoerd TC, van den Elzen PJM, Hoekema A. Production of correctly processed human serum albumin in transgenic plants. Bio/Technology. 1990;**83**(8):217-221. DOI: 10.1038/nbt0390-217

[32] Pipeline—iBio. n.d. Available from: https://www.ibioinc.com/pipeline/ [Accessed: March 18, 2021]

[33] Medicago R&D Pipeline: Influenza, Rotevirus, Norovirus and Coronavirus. Available from: https://medicago.com/en/our-science/our-pipeline/ [Accessed: March 18, 2021]

[34] Icon Genetics. Available from: https://www.icongenetics.com/ [Accessed: March 18, 2021]

[35] Shaaltiel Y, Bartfeld D, Hashmueli S, Baum G, Brill-Almon E, Galili G, et al. Production of glucocerebrosidase with terminal mannose glycans for enzyme replacement therapy of Gaucher's disease using a plant cell system. Plant Biotechnology Journal. 2007;**5**:579-590. DOI: 10.1111/j.1467-7652.2007.00263.x

[36] Pharmaceutical Research And Development | Wuhan Healthgen Biotechnology Corp. Available from: https://www.oryzogen.net [Accessed: March 18, 2021]

[37] Planet Biotechnology, Inc. Available from: https://www.planetbiotechnology.com [Accessed: March 18, 2021]

[38] Lactoferrin human recombinant, expressed in rice, Iron saturated, 90 (SDS-PAGE) Holo Transferrin. Available from: https://www.sigmaaldrich.com/TR/en/product/sigma/l1294 [Accessed: March 18, 2021]

[39] Interleukin-12 p40 Human Recombinant, Plant. Available from: https://www.angioproteomie.com/commerce/ccp10634-interleukin-12-p40-human-recombinant--plant-rap-0487.htm [Accessed: March 18, 2021]

[40] Abbexa Ltd. Available from: https://www.abbexa.com/fibroblast-growth-factor-basic-plant-protein-recombinant [Accessed: March 18, 2021]

[41] Hudson LC, Garg R, Bost KL, Piller KJ. Soybean seeds: a practical host for the production of functional subunit vaccines. BioMed Research International. 2014;**2014**. DOI: 10.1155/2014/340804

[42] Maharjan PM, Cheon J, Jung J, Kim H, Lee J, Song M, et al. Plant-expressed receptor binding domain of the SARS-CoV-2 spike protein elicits humoral immunity in mice. Vaccines. 2021;**9**:978. DOI: 10.3390/VACCINES9090978

[43] Our Discovery Pipeline—KBP. (n.d.)

[44] Ventria Bioscience. Available from: https://ventria.com/medicines/infectious-diseases/venbeta [Accessed: March 18, 2021]

[45] Thermo Fisher Scientific Inc. Available from: https://www.thermofisher.com/order/catalog/product/A35934 [Accessed: March 18, 2021]

[46] ORF Genetics. Available from: https://www.orfgenetics.com/product/isokine [Accessed: March 18, 2021]

[47] Spiegel H, Stöger E, Twyman RM, Buyel JF. Current status and perspectives of the molecular farming landscape. Molecular Pharming: Applications, Challenges and Emerging Areas. 2018:1-23. DOI: 10.1002/9781118801512.CH1

[48] Düring K, Hippe S, Kreuzaler F, Schell J. Molecular biology, Undefined 1990, Synthesis and Self-Assembly of a Functional Monoclonal Antibody in Transgenic Nicotiana tabacum. Berlin: Springer; n.d.

[49] J. Ma, A. Hiatt, M. Hein, N. Vine, Generation and Assembly of Secretory Antibodies in Plants, Science. Sciencemag.Org. (n.d.). Available from: https://science.sciencemag.org/content/268/5211/716.abstract [Accessed: November 24, 2021]

[50] Phakham T, Bulaon CJI, Khorattanakulchai N, Shanmugaraj B, Buranapraditkun S, Boonkrai C, et al. Functional characterization of pembrolizumab produced in nicotiana benthamiana using a rapid transient expression system. Frontiers in Plant Science. 2021;**12**. DOI: 10.3389/FPLS.2021.736299

[51] Hurtado J, Acharya D, Lai H, Sun H, Kallolimath S, Steinkellner H, et al. In vitro and in vivo efficacy of anti-chikungunya virus monoclonal antibodies produced in wild-type and glycoengineered Nicotiana benthamiana plants. Plant Biotechnology Journal. 2020;**18**:266-273. DOI: 10.1111/PBI.13194

[52] Rattanapisit K, Shanmugaraj B, Manopwisedjaroen S, Purwono PB, Siriwattananon K, Khorattanakulchai N, et al. Rapid production of SARS-CoV-2 receptor binding domain (RBD) and spike specific monoclonal antibody CR3022 in Nicotiana benthamiana. Scientific Reports. 2020;**10**:1-11. DOI: 10.1038/s41598-020-74904-1

[53] Diamos AG, Hunter JGL, Pardhe MD, Rosenthal SH, Sun H, Foster BC, et al. High level production of monoclonal antibodies using an optimized plant expression system. Frontiers in Bioengineering and Biotechnology. 2020. Article ID: 472. DOI: 10.3389/FBIOE.2019.00472

[54] Rattanapisit K, Chao Z, Siriwattananon K, Huang Z, Phoolcharoen W. Plant-produced anti-enterovirus 71 (EV71) monoclonal antibody efficiently protects mice against EV71 infection. Plants. 2019;**8**. DOI: 10.3390/PLANTS8120560

[55] Choi YS, Moon JH, Kim TG, Lee JY. Potent in vitro and in vivo activity of plantibody specific for porphyromonas gingivalis FimA. Clinical and Vaccine Immunology. 2016;**23**:346-352. DOI: 10.1128/CVI.00620-15

[56] Dent M, Hurtado J, Paul AM, Sun H, Lai H, Yang M, et al. Plant-produced anti-dengue virus monoclonal antibodies exhibit reduced antibody-dependent enhancement of infection activity. The Journal of General Virology. 2016;**97**:3280. DOI: 10.1099/JGV.0.000635

[57] Buyel JF, Fischer R. Scale-down models to optimize a filter train for the downstream purification of recombinant pharmaceutical proteins produced in tobacco leaves. Biotechnology Journal. 2014;**9**:415-425. DOI: 10.1002/BIOT.201300369

[58] Ma JK, Drossard J, Lewis D, Altmann F, Boyle J, Christou P, et al. Regulatory approval and a first-in-human phase I clinical trial of a monoclonal antibody produced in

transgenic tobacco plants. Plant Biotechnology Journal. 2015;**13**: 1106-1120. DOI: 10.1111/PBI.12416

[59] Sabalza M, Madeira L, van Dolleweerd C, Ma JK, Capell T, Christou P. Functional characterization of the recombinant HIV-neutralizing monoclonal antibody 2F5 produced in maize seeds. Plant Molecular Biology. 2012;**804**(80):477-488. DOI: 10.1007/S11103-012-9962-6

[60] Drake PMW, Barbi T, Sexton A, McGowan E, Stadlmann J, Navarre C, et al. Development of rhizosecretion as a production system for recombinant proteins from hydroponic cultivated tobacco. The FASEB Journal. 2009;**23**: 3581-3589. DOI: 10.1096/FJ.09-131771

[61] Lai H, He J, Hurtado J, Stahnke J, Fuchs A, Mehlhop E, et al. Structural and functional characterization of an anti-West Nile virus monoclonal antibody and its single-chain variant produced in glycoengineered plants. Plant Biotechnology Journal. 2014;**12**: 1098-1107. DOI: 10.1111/pbi.12217

[62] Hernández-Velázquez A, López-Quesada A, Ceballo-Cámara Y, Cabrera-Herrera G, Tiel-González K, Mirabal-Ortega L, et al. Tobacco seeds as efficient production platform for a biologically active anti-HBsAg monoclonal antibody. Transgenic Research. 2015;**24**:897-909. DOI: 10.1007/S11248-015-9890-8

[63] van Dolleweerd CJ, Teh AY-H, Banyard AC, Both L, Lotter-Stark HCT, Tsekoa T, et al. Engineering, Expression in Transgenic Plants and Characterisation of E559, a Rabies Virus-Neutralising Monoclonal Antibody. The Journal of Infectious Diseases. 2014;**210**:200-208. DOI: 10.1093/INFDIS/JIU085

[64] Lai H, He J, Engle M, Diamond MS, Chen Q. Robust production of virus-like particles and monoclonal antibodies with geminiviral replicon vectors in lettuce. Plant Biotechnology Journal. 2012;**10**:95-104. DOI: 10.1111/J.1467-7652.2011.00649.X

[65] Brodzik R, Glogowska M, Bandurska K, Okulicz M, Deka D, Ko K, et al. Plant-derived anti-Lewis Y mAb exhibits biological activities for efficient immunotherapy against human cancer cells. Proceedings of the National Academy of Sciences. 2006;**103**(23): 8804-8809

[66] Nakanishi K, Morikane S, Hosokawa N, Kajihara Y, Kurohane K, Niwa Y, et al. Plant-derived secretory component forms secretory IgA with shiga toxin 1-specific dimeric IgA produced by mouse cells and whole plants. Plant Cell Reports. 2019;**38**:161-172. DOI: 10.1007/S00299-018-2358-6

[67] Nakanishi K, Morikane S, Ichikawa S, Kurohane K, Niwa Y, Akimoto Y, et al. Protection of human colon cells from Shiga toxin by plant-based recombinant secretory IgA. Nature.Com. n.d.

[68] Posgai AL, Wasserfall CH, Kwon KC, Daniell H, Schatz DA, Atkinson MA. Plant-based vaccines for oral delivery of type 1 diabetes-related autoantigens: Evaluating oral tolerance mechanisms and disease prevention in NOD mice. Scientific Reports. 2017;7: 1-15. DOI: 10.1038/srep42372

[69] Bertini E, Merlin M, Gecchele E, Puggia A, Brozzetti A, Commisso M, et al. Design of a type-1 diabetes vaccine candidate using edible plants expressing a major autoantigen. Frontiers in Plant Science. 2018;**9**:1-16. DOI: 10.3389/fpls.2018.00572

[70] Arevalo-Villalobos JI, Govea-Alonso DO, Monreal-Escalante E, Zarazúa S, Rosales-Mendoza S. LTB-Syn: A recombinant immunogen for the development of plant-made vaccines against synucleinopathies. Planta.

2017;**245**:1231-1239. DOI: 10.1007/s00425-017-2675-y

[71] Suzuki K, Kaminuma O, Yang L, Takai T, Mori A, Umezu-Goto M, et al. Prevention of allergic asthma by vaccination with transgenic rice seed expressing mite allergen: Induction of allergen-specific oral tolerance without bystander suppression. Plant Biotechnology Journal. 2011;**9**:982-990. DOI: 10.1111/j.1467-7652.2011.00613.x

[72] Youm JW, Jeon JH, Kim H, Kim YH, Ko K, Joung H, et al. Transgenic tomatoes expressing human beta-amyloid for use as a vaccine against Alzheimer's disease. Biotechnology Letters. 2008;**30**:1839-1845. DOI: 10.1007/s10529-008-9759-5

[73] Takagi H, Hiroi T, Yang L, Tada Y, Yuki Y, Takamura K, et al. A rice-based edible vaccine expressing multiple T cell epitopes induces oral tolerance for inhibition of Th2-mediated IgE responses. Proceedings of the National Academy of Sciences of the United States of America. 2005;**102**:17525-17530. DOI: 10.1073/pnas.0503428102

[74] Zheng N, Xia R, Yang C, Yin B, Li Y, Duan C, et al. Boosted expression of the SARS-CoV nucleocapsid protein in tobacco and its immunogenicity in mice. Vaccine. 2009;**27**:5001-5007. DOI: 10.1016/j.vaccine.2009.05.073

[75] Monreal-Escalante E, Ramos-Vega AA, Salazar-González JA, Bañuelos-Hernández B, Angulo C, Rosales-Mendoza S. Expression of the VP40 antigen from the Zaire ebolavirus in tobacco plants. Planta. 2017;**246**:123-132. DOI: 10.1007/s00425-017-2689-5

[76] Yang M, Sun H, Lai H, Hurtado J, Chen Q. Plant-produced Zika virus envelope protein elicits neutralizing immune responses that correlate with protective immunity against Zika virus in mice. Plant Biotechnology Journal. 2018;**16**:572-580. DOI: 10.1111/pbi.12796

[77] Kashima K, Yuki Y, Mejima M, Kurokawa S, Suzuki Y, Minakawa S, et al. Good Manufacturing Practices Production of a Purification-Free Oral Cholera Vaccine Expressed in Transgenic Rice Plants. Berlin: Springer; 1911. DOI: 10.1007/s00299-015-1911-9

[78] Makarkov AI, Golizeh M, Ruiz-Lancheros E, Gopal AA, Costas-Cancelas IN, Chierzi S, et al. Plant-derived virus-like particle vaccines drive cross-presentation of influenza A hemagglutinin peptides by human monocyte-derived macrophages. Npj Vaccines. 2019;**4**. DOI: 10.1038/s41541-019-0111-y

[79] Mbewana S, Mortimer E, Pêra FFPG, Hitzeroth II, Rybicki EP. Production of H5N1 influenza virus matrix protein 2 ectodomain protein bodies in tobacco plants and in insect cells as a candidate universal influenza vaccine. Frontiers in Bioengineering and Biotechnology. 2015;**3**:1-8. DOI: 10.3389/fbioe.2015.00197

[80] Alvarez ML, Pinyerd HL, Crisantes JD, Rigano MM, Pinkhasov J, Walmsley AM, et al. Plant-made subunit vaccine against pneumonic and bubonic plague is orally immunogenic in mice. Vaccine. 2006;**24**:2477-2490. DOI: 10.1016/j.vaccine.2005.12.057

[81] Tottey S, Shoji Y, Jones RM, Chichester JA, Green BJ, Musiychuk K, et al. Plant-produced subunit vaccine candidates against yellow fever induce virus neutralizing antibodies and confer protection against viral challenge in animal models. The American Journal of Tropical Medicine and Hygiene. 2018;**98**:420-431. DOI: 10.4269/ajtmh.16-0293

[82] Huang Y, Yang C, Xu X, Xu W, Sinica SL-AP. Structural and functional properties of SARS-CoV-2 spike protein: potential antivirus drug development for COVID-19. Nature.Com. n.d. Available from: https://www.nature.

com/articles/s41401-020-0485-4?elqTrackId=75b9ed6bfda8445ba6e819d53c47c494 [Accessed: November 16, 2021

[83] Dai L, Gao GF. Viral targets for vaccines against COVID-19. Nature Reviews Immunology. 2020;**212**(21):73-82. DOI: 10.1038/s41577-020-00480-0

[84] Dhama K, Natesan S, Iqbal Yatoo M, Kumar Patel S, Tiwari R, Saxena SK, et al. Plant-based vaccines and antibodies to combat COVID-19: current status and prospects. Taylor and Francis. 2020;**16**: 2913-2920. DOI: 10.1080/21645515.2020.1842034

[85] Ward BJ, Gobeil P, Séguin A, Atkins J, Boulay I, Charbonneau PY, et al. Phase 1 randomized trial of a plant-derived virus-like particle vaccine for COVID-19. Nature Medicine. 2021; **276**(27):1071-1078. DOI: 10.1038/s41591-021-01370-1

[86] KBP-201 COVID-19 Vaccine Trial in Healthy Volunteers – Full Text View – ClinicalTrials.gov. (n.d.). Available from: https://clinicaltrials.gov/ct2/show/NCT04473690?term=KBP+201&draw=1&rank=1 [Accessed: November 17, 2021]

[87] COVID-19—Baiya Phytopharm. n.d.. Available from: https://baiyaphytopharm.com/covid-19/ [Accessed: November 17, 2021]

[88] Barta A, Sommergruber K, Thompson D, Hartmuth K, Matzke MA, Matzke AJM. The expression of a nopaline synthase—Human growth hormone chimaeric gene in transformed tobacco and sunflower callus tissue. Plant Molecular Biology. 1986;**65**(6): 347-357. DOI: 10.1007/BF00034942

[89] Ahmad AR, Kaewpungsup P, Khorattanakulchai N, Rattanapisit K, Pavasant P, Phoolcharoen W. Recombinant human dentin matrix protein 1 (HDMP1) expressed in nicotiana benthamiana potentially inducesosteogenic differentiation. Plants. 2019;**8**:1-12. DOI: 10.3390/plants8120566

[90] Mbewana S, Meyers AE, Weber B, Mareledwane V, Ferreira ML, Majiwa PAO, et al. Expression of rift valley fever virus N-protein in Nicotiana benthamiana for use as a diagnostic antigen. BMC Biotechnology. 2018;**18**:1-10. DOI: 10.1186/s12896-018-0489-z

[91] Gurusamy PD, Schäfer H, Ramamoorthy S, Wink M. Biologically active recombinant human erythropoietin expressed in hairy root cultures and regenerated plantlets of Nicotiana tabacum L. PLoS One. 2017;**12**:e0182367. DOI: 10.1371/JOURNAL.PONE.0182367

[92] Rattanapisit K, Abdulheem S, Chaikeawkaew D, Kubera A, Mason HS, Ma JK-C, et al. Recombinant human osteopontin expressed in Nicotiana benthamiana stimulates osteogenesis related genes in human periodontal ligament cells. Scitific Reports. 2017;**7**: 1-9. DOI: 10.1038/s41598-017-17666-7

[93] Shen J-S, Busch A, Day TS, Meng X-L, Yu CI, Dabrowska-Schlepp P, et al. Mannose receptor-mediated delivery of moss-made α-galactosidase A efficiently corrects enzyme deficiency in Fabry mice. Journal of Inherited Metabolic Disease. 2016;**39**:293. DOI: 10.1007/S10545-015-9886-9

[94] Limkul J, Misaki R, Kato K, Fujiyama K. The combination of plant translational enhancers and terminator increase the expression of human glucocerebrosidase in Nicotiana benthamiana plants. Plant Science. 2015;**240**:41-49. DOI: 10.1016/j.plantsci.2015.08.018

[95] Su J, Sherman A, Doerfler PA, Byrne BJ, Herzog RW, Daniell H. Oral delivery of Acid Alpha Glucosidase epitopes expressed in plant chloroplasts suppresses antibody formation in treatment of Pompe mice. Plant

Biotechnology Journal. 2015;**13**: 1023-1032. DOI: 10.1111/PBI.12413

[96] Tan Y, Wang KY, Wang N, Li G, Liu D. Ectopic expression of human acidic fibroblast growth factor 1 in the medicinal plant, Salvia miltiorrhiza, accelerates the healing of burn wounds. BMC Biotechnology. 2014;**14**:1-10. DOI: 10.1186/1472-6750-14-74

[97] Guan C, Du X, Wang G, Ji J, Jin C, Li X. Expression of biologically active anti-thrombosis protein lumbrokinase in edible sunflower seed kernel. Journal of Plant Biochemistry and Biotechnology. 2014;**23**:257-265. DOI: 10.1007/s13562-013-0209-7

[98] Reuter LJ, Bailey MJ, Joensuu JJ, Ritala A. Scale-up of hydrophobin-assisted recombinant protein production in tobacco BY-2 suspension cells. Plant Biotechnology Journal. 2014;**12**:402-410. DOI: 10.1111/PBI.12147

[99] Arzola L, Chen J, Rattanaporn K, Maclean JM, McDonald KA. Transient co-expression of post-transcriptional gene silencing suppressors for increased in planta expression of a recombinant anthrax receptor fusion protein. International Journal of Molecular Sciences. 2011;**12**:4975. DOI: 10.3390/IJMS12084975

[100] Pogue GP, Vojdani F, Palmer KE, Hiatt E, Hume S, Phelps J, et al. Production of pharmaceutical-grade recombinant aprotinin and a monoclonal antibody product using plant-based transient expression systems. Plant Biotechnology Journal. 2010;**8**:638-654. DOI: 10.1111/J.1467-7652.2009.00495.X

[101] Kim TG, Baek MY, Lee EK, Kwon TH, Yang MS. Expression of human growth hormone in transgenic rice cell suspension culture. Plant Cell Reports. 2008;**27**:885-891. DOI: 10.1007/s00299-008-0514-0

[102] Ding SH, Huang LY, Wang YD, Sun HC, Xiang ZH. High-level expression of basic fibroblast growth factor in transgenic soybean seeds and characterization of its biological activity. Biotechnology Letters. 2006;**28**:869-875. DOI: 10.1007/s10529-006-9018-6

[103] Komarnytsky S, Borisjuk NV, Borisjuk LG, Alam MZ, Raskin I. Production of recombinant proteins in tobacco guttation fluid. Plant Physiology. 2000;**124**:927-934. DOI: 10.1104/PP.124.3.927

[104] D. Yang, et al. Transgenic Rice for the Production of Recombinant Pharmaceutical Proteins: A Case Study of Human Serum Albumin, Books. Google.Com. n.d. Available from: https://www.google.com/books?hl=en&lr=&id=KwNRDwAAQBAJ&oi=fnd&pg=PA275&dq=Transgenic+Rice+for+the+Production+of+Recombinant+Pharmaceutical+Proteins:+A+Case+Study+of+Human&ots=JPkKjYIN8C&sig=Ujk2YrqZsWoWR_hwMtn2GIinCmI [Accessed: November 17, 2021]

[105] Sedaghati B, Haddad R, Bandehpour M. Transient expression of human serum albumin (HSA) in tobacco leaves. Molecular Biology Reports. 2020;**47**:7169-7177. DOI: 10.1007/S11033-020-05640-Y

[106] Zhu Z, Hughes KW, Huang L, Sun B, Liu C, Li Y. Expression of human α-interferon cDNA in transgenic rice plants. Plant Cell, Tissue and Organ Culture. 1994;**36**:197-204. DOI: 10.1007/BF00037720

[107] Matsumoto S, Ikura K, Ueda M, Sasaki R. Characterization of a human glycoprotein (erythropoietin) produced in cultured tobacco cells. Plant Molecular Biology. 1995;**27**:1163-1172. DOI: 10.1007/BF00020889

[108] He M, Kittur FS, Hung CY, Zhang J, Jing L, Sane DC, et al. A novel

plant-produced asialo-rhuepo protects brain from ischemic damage without erythropoietic action. Transl Stroke Res. 2021 Sep 22. DOI: 10.1007/s12975-021-00943-z. [Epub ahead of print]. PMID: 34553324

[109] Ruggiero F, Exposito JY, Bournat P, Gruber V, Perret S, Comte J, et al. Triple helix assembly and processing of human collagen produced in transgenic tobacco plants. FEBS Letters. 2000;**469**:132-136. DOI: 10.1016/S0014-5793(00)01259-X

[110] Merle C, Perret S, Lacour T, Jonval V, Hudaverdian S, Garrone R, et al. Hydroxylated human homotrimeric collagen I in Agrobacterium tumefaciens-mediated transient expression and in transgenic tobacco plant. FEBS Letters. 2002;**515**:114-118. DOI: 10.1016/S0014-5793(02)02452-3

Chapter 4

Site-Specific Recombination and the Development of Cisgenic Plants

Richard Mundembe

Abstract

The commercialization of transgenic plants almost three decades ago was accompanied by controversies that highlighted concerns that relate to human health and the environment. This has resulted in continued research efforts to further improve molecular genetic approaches to plant genetic engineering. One such approach involves the use of site-specific recombination mechanisms to produce cisgenic plants. This chapter describes the different methods for site-specific recombination and briefly comments on their potential for widespread adoption in the production of cisgenic plants. The chapter concludes by showcasing some cisgenic plants under development and highlights how cisgenic plants circumvent some concerns associated with first-generation transgenic plants.

Keywords: transgenic plants, cisgenic plants, site-specific recombination, genome-editing

1. Introduction

Transgenic plants have been around for some time and have become of age. Their strict regulation and public approval processes remain controversial with some people opposed to transgenic plants for reasons that can broadly be categorized into real, perceived or imagined based on established scientific facts. During evaluations before releasing into the environment, the transgenic plants are evaluated to confirm if they deliver the benefits they claim, and whether they are safe to human health and the environment [1]. Continued engagement with these concerns has led to the exploration of possible solutions that make the technology more efficient, safer, and more widely accepted.

Many of the concerns associated with genetically modified plants emanate from the use of reporter genes and selectable marker genes such as antibiotic or herbicide resistance genes in the process of genetic engineering [2, 3]. These genes, together with the transgene of interest, are usually part of a relatively large plasmid that has sequences that are not required in the genetically modified plant but become transferred to the plant and get integrated into the plant genome because of the general and non-specific nature of plant transformation methods. The reporter and selectable marker genes together with vector backbone sequences once present in the transgenic plant are of ecological concern because they may be passed on to other species in the environment, resulting in such characteristics as weediness and

IntechOpen

invasiveness [1, 2]. The spread of antibiotic resistance genes is also of great concern to human health. The integration sites of these genes are random and may result in some 'unintended effects' such as inactivation of important genes and production of new toxins or allergens. All these factors are considered during evaluations for the release of genetically modified plants.

Over the years, the concept of cisgenic plants has emerged and is contrasted with transgenic plants by using DNA sequences from naturally crossable species and possibly avoiding the use of reporter and selectable marker genes as well as vector sequences [4, 5]. This chapter will explain how that is achieved, and why cisgenic plants might be more widely acceptable to regulators and consumers.

2. History and controversies of transgenics/GMOs

In 1994, the genetically modified Flavr Savr™ tomato was commercialized [6]. This was a great stride for both science and commerce, and the society's response to this new type of product has helped determine how such new technologies are regulated. While some proponents of genetic engineering would have wished for no labelling and minimum statutory regulation of the development and environmental release of the GMOs, an antagonistic anti-GMO movement arose and advocated for a 'ban' on GMOs. This created a healthy, restrained environment in which real, potential and imagined dangers of the new technology could be objectively evaluated. Processes for approval and release of GMOs were established, enabling society to have a say, whatever the nature of their reservations might be.

Singh et al. [1] lists five potential risks associated with the cultivation of transgenic crops. These are: (1) Introduction of allergenic or harmful proteins into the foods; (2) Detrimental effects on non-target species and the environment; (3) Increased invasiveness and weediness of crop plants; (4) Increase pest and disease resistance in response to intense selection pressure; and (5) Fear of biodiversity loss. These potential risks must be addressed before regulatory approval for the release of a transgenic plant is granted. Scientific research has therefore continued to look for ways to eliminate the sources of these concerns, where possible.

Many of the potential risks related to the presence of a reporter and selectable marker genes in the GMOs. Vector backbone sequences often get integrated into the plant genome as well [2]. The sites of integration of these DNA sequences are often random, possibly disrupting some essential gene functions, giving rise to toxic or allergenic products and some other non-intended effects [5]. Alternative methods of genetically engineering plants have been developed to better address some of these concerns.

Two main methods are used for plant transformation: *Agrobacterium*-mediated transformation and biolistics (bombardment) [2, 3]. The processes have been studied for a long time and there is some understanding of how transgenes enter the cell cytoplasm and nucleus in both cases, the mechanisms for transgene integration into the genome cannot be easily manipulated. The mechanisms of integration involve homologous or non-homologous recombination and are reviewed by Mundembe and Hwang et al. [7, 8].

It became apparent that the methods of plant transformation and mechanism of DNA integration were intricately linked to the concerns raised against transgenic plants [2]. The selectable marker such as herbicide resistance or antibiotic resistance gene is required as a mechanism to positively select for transformed plants over untransformed plants; untransformed plants will not survive in the presence of the herbicide or antibiotic [2]. The reporter gene gives a visual marker such as colour or fluorescence that enables the experimenter to tell the transformed nature of any tissue easily. These marker genes are usually on the same piece of DNA as the gene of interest so that the presence of the marker genes can be taken as an indication of the presence of the gene of interest as well. The presence of these genes in the environment is a major concern. In addition, the site of integration of these genes is random.

Vector backbone sequences also often become integrated into the plant genome. Applicants for approval are required to demonstrate that vector backbone sequences are absent for approval to be granted. Plant transformation experiments are designed on the assumption that only the sequences between the left and right borders of a T-DNA will be transferred to the plant genome. Widespread reports of integration of vector sequences were cited by opponents of genetic engineering as evidence that the genetic engineering of plants was not sufficiently understood to be released into the environment. The perception of 'randomness' of transfer and integration made the public uneasy about GMOs.

Site-specific recombination promised to circumvent the concerns about the randomness of the site of integration. Site-specific recombination systems have been studied since the 1980s. These include Cre-lox P ('*c*auses/*c*yclization *r*ecombination/*lo*cus of crossing over, *x*, in P1'), FLP/FRP (flippase/flippase recognition target) and λ integrase [9]. Zinc finger nucleases (ZFN), transcription activator-like effector nucleases (TALENs) and clustered regularly interspaced short palindromic repeats- CRISPR-associated gene 9 (CRISPR-Cas9) is another group of nucleases that has been adapted for manipulation of DNA at specific sites [10].

3. Recombination

Recombination is defined as the production of new DNA molecule(s) from two parental DNA molecules or different segments of the same DNA molecule [11]. Recombination is essential for the success of all living organisms, to generate diversity and to repair damage to DNA. The types of recombination include homologous (general) recombination, nonhomologous (illegitimate) recombination, replicative recombination (transposition) and site-specific recombination.

3.1 Homologous recombination

Homologous recombination occurs throughout genomes of diploid organisms immediately before cell division in the late S or G_2 phases of the cell cycle and is responsible for recombining large pieces of DNA that have a very similar sequences [12]. The mechanism of homologous recombination is complex and may involve many enzymes but is very accurate and tightly controlled. It can repair double-stranded breaks with either single or double ends, even those with covalently attached proteins [13]. Holliday junctions are formed, and their resolution determines the outcome.

3.2 Nonhomologous recombination

Nonhomologous (illegitimate) recombination occurs in regions where no large-scale sequence similarity is apparent and is responsible for translocations between nonhomologous chromosomes or deletions of several genes from a chromosome [13]. It is the main mechanism for DNA repair that takes place throughout the cell cycle, repairing DNA damage due to chemicals and UV light. It efficiently restores chromosomal integrity at the risk of introducing local sequence errors.

The mechanisms of non-homologous recombination are nonhomologous end-joining (NHEJ) and alternative NHEJ (altNHEJ, also known as microhomology-mediated end-joining, MMEJ). They involve the ligation of two double-stranded breaks with little or no sequence homology, without the need for a repair template [13, 14].

3.3 Replicative recombination

This is a specialized type of recombination in which a segment of DNA is translocated from one location on a chromosome to another on the same or another chromosome in a process that involves the generation of a new copy of a segment of DNA [11]. Many transposable genetic elements use this process to generate a new copy of the transposable genetic element at a new location.

3.4 Site-specific recombination

Site-specific recombination (SSR) is widespread in prokaryotes, involves much shorter DNA segments and requires specific nucleotide sequences that are recognized by specific proteins known as recombinases. The lambda integrase system for integration into *E. coli* genome was the first to be discovered, but many more systems have since been discovered and characterized. Site-specific recombination brings together two short DNA sequences on separate locations on the same or separate DNA molecules, with the cutting and re-joining of the DNA molecules in a recombination reaction catalyzed by specific SSR enzyme systems [10, 15]. The process is conservative since it does not involve DNA synthesis or degradation, or any high-energy cofactors such as ATP, and is thus distinct from homologous, non-homologous and replicative recombination. The outcomes of SSR are integration/excision, inversion or linear recombination depending on the initial orientation of the two target sites.

The conservative site-specific recombinases can be classified into two families: serine family recombinases (formerly known as invertase/resolvase) and tyrosine family recombinases (formerly known as integrase) based on the amino acid that acts as the active site nucleophile during DNA breakage [15]. An example of serine family recombinase is bacteriophage PhiC31 integrase. Examples of tyrosine family recombinases are Lambda integrase, Cre recombinase and Flp recombinase.

The serine family recombinases carry out DNA inversion or DNA resolution (excision) reactions. The mechanism involves staggered double-stranded breaks in two parallel dsDNA molecules participating in the exchange, followed by a 180° rotation of the recombination complex (in a plane perpendicular to that of the DNA molecules), and then ligated. The tyrosine family recombinases carry out DNA integration reactions. The mechanism involves the formation of a Holliday junction because of initial cuts in only one (inner) strand of each of two dsDNA molecules positioned antiparallel to each other, and they are rejoining across the molecules. The Holliday junction is resolved when the outer DNA molecules are also cut and rejoined to result in recombinant DNA strands [15]. The reader is referred to Jayaram et al. [15] for more details of recombination geometries.

In-plant biotechnology, the cre-loxP system is a historically prominent SSR system and will be considered in more detail below. Recently, the CRISPR-Cas 9 system and related nuclease variants have gained great prominence and will also be considered in detail.

3.4.1 Cre-loxP recombination system

The Cre-*lox*P site-specific recombination system is based on a naturally occurring Bacteriophage P1 system. The name 'cre' is derived from '*c*auses/*c*yclization *r*ecombination' while '*lox*P' is derived from '*lo*cus of crossing over, *x*, in P1'. The *lox*P site is composed of 34 bp consensus sequence consisting of an 8 bp nonsymmetrical central region flanked by two 13 bp palindromic sequences. Cre recombinase is a 38 kDa protein that catalyses the recombination of two *lox*P recognition sites on the same or different DNA strands using tyrosine 324 for the nucleophilic attack [9]. The recombination takes place via a Holliday junction intermediate formed by two antiparallel DNA molecules/segments to which a dimer of Cre recombinase subunits is bound to each *lox*P site. Two opposite active Cre recombinase subunits catalyse strand cleavage, exchange and ligation at the 8 bp nonsymmetric central region, thus resolving the Holliday junction intermediate. Excision of DNA flanking two *lox*P sequences occurs if the two have the same orientation; if their orientation is opposite, then inversion of the intervening sequence occurs. Strand exchange or translocation will occur if two *lox*P sites located on different DNA molecules recombine.

3.4.2 CRISPR/Cas9 and other variants

The CRISPR system was first reported in 1987 in *E. coli* where it functions as a form of adaptive immunity against invading nucleic acid [16] and has since been shown to be of ubiquitous occurrence [17]. Many variations have since been discovered in nature, and modifications have also been introduced by genetic engineering for ease of use.

The CRISPR/Cas9 system currently used is composed of an RNA-dependent DNA endonuclease called Cas 9 protein, complexed with a guide RNA (gRNA). The gRNA is only 20 nucleotides long and is complementary to the target DNA to which it recruits the Cas9 protein [18]. The Cas9/gRNA then binds to a short but specific protospacer adjacent motif (PAM) sequence at the 3′ end of the target sequence. For *Streptococcus pyogenes*, the PAM sequence is 5'-NGG-3′. Cas9 protein then introduces a double-stranded break (DSB) on the target sequence. The DSB will be repaired by HR or NHEJ, resulting in insertion, deletion, or fragment replacement within the target site. Thus, recombination will be effected.

4. Cisgenics

Cisgenesis is defined as the genetic modification of a recipient plant with a natural gene (in the sense orientation, with its natural promoter, terminator and introns) from a sexually compatible plant [5]. A closely related concept is that of intragenesis, where an additional hybrid copy of a gene from the same or crossable species is introduced in sense or anti-sense orientation, combining promoter, coding region and terminator from different genes [19–22]. In intragenesis therefore, some changes or reshuffling of coding or control regions of the natural gene(s) will have occurred, unlike in cisgenesis. In addition, Rommens et al. [4] stipulates that for *Agrobacterium*-mediated transformation, border sequences derived from plants (P-DNA) should be used in place of T-DNA. Cisgenesis and intragenesis are contrasted with transgenesis, which is the genetic modification of a recipient plant with one or more genes from any non-plant organism, or from a donor plant that is sexually incompatible with the recipient plant. Holme et al. [21] discusses the varying stringency with which the term 'cisgenic' has been used over the years. The strictest definitions of the terms are advocated for since technological advances now enable more precise genetic modification followed by more detailed sequence analysis of the resulting genetically modified plants. This would also facilitate the implementation of different regulatory regimes for cisgenic and transgenic plants.

Early definitions of cisgenesis emphasized the source of the gene of interest used in transformation and may not have insisted on the complete absence of other accompanying sequences. At that time, almost all transgenic plants were developed using *Agrobacterium*-mediated transformation or biolistics, with the gene of interest being introduced as part of a binary plasmid with the reporter and selectable marker genes. The least stringent definition did not fully consider the possible presence of these extra genetic sequences, and the sites of insertion. Later reports of cisgenesis included procedures to remove extra sequences via traditional crossing or by site-specific recombination procedures.

The strictest definition of cisgenesis should apply only when the procedures through which the plant was modified do not involve any DNA sequences, however short or procedurally essential, from any non-plant organism or sexually incompatible plant. This strict definition has recently become possible because of recently developed tools for site-specific recombination and genome editing. Some examples are considered below.

Many different strategies have been used to meet the marker-free status that is required for cisgenic plants. Where transformation efficiencies are high, plant transformation can be carried out using constructs that do not have selectable markers; transformed lines are identified by screening for the specific gene sequences that have been introduced. Biolistic transformation using appropriate minimal cassettes has also been suggested [7]. This requires analysis of many lines, which makes it time-consuming and expensive.

In an alternative strategy, constructs in which selectable markers are flanked by site-specific recombination sites have been used. The selectable markers are later deleted from transformed plants following induction of the site-specific recombination system. Examples, where this approach was used, are in intragenic strawberries [23] and in cisgenic and intragenic apples [24, 25]. In maize, a series of transgenic maize lines that express five different recombinases have been generated and can be used for selectable marker removal and transgene integration into specific loci [26].

Marker-free transformants may also be obtained through a co-transformation strategy, where the selectable marker and the transgene of interest are introduced on different vector constructs so that they integrate into different locations on the plant genome. The two genes may then segregate into different progeny in subsequent generations. Cisgenic durum wheat [27] and cisgenic barley [28] were generated using this strategy.

In all these strategies however the site of integration of the transgene is random, and there is always a chance that vector backbone sequences may also be integrated into the plant. Recent work with CRISPR-based strategies has attempted to address these shortcomings.

5. Genome-editing technologies and cisgenics

Genome editing is the addition, removal or alteration of genetic material (at particular locations) in the genome of an organism. Concurrent developments in site-specific recombination and genome sequencing technologies have made (precision) genome editing a reality. It is now possible to sequence the whole genome of an organism in a very short period and at a cost that is affordable to research laboratories. Many site-specific recombination systems have been developed into technologies that can target specific sites in the genome at which specific, pre-determined changes will be introduced. Re-sequencing of the genome will verify the specificity of the modifications.

To initiate genome-editing, double-stranded breaks are made in the target genome at the site to be edited. Many tools have been developed for precision targeting of these double-stranded breaks. These include meganucleases, zinc finger nucleases (ZFN), transcription activator-like effector nucleases (TALENS) and CRISPR/Cas. The requirements for recognition of specific DNA nucleotide sequence sites and mechanisms of cleavage for these nucleases are reviewed in detail elsewhere [10]. Once the double-stranded cuts have been made on a DNA molecule, endogenous cellular factors recognise and bind to these sites of discontinuity and initiate repair by either HR or NHEJ mechanisms resulting in addition, removal or some other kind of alteration of the DNA nucleotide sequence following the design of the editing system used. In this paper, we will use the CRISPR/Cas system to further illustrate this, and show how cisgenic plants *sensus stricto* can be obtained.

Truly cisgenic plants should be a reporter- and selection marker-free, should not contain sequences from non-crossable species, and the editing must be done by a precise mechanism at a pre-determined genomic site. Most of the reports on genome editing do not result in cisgenic plants because they do not satisfy at least one of these requirements. Most reports use selection marker genes, DNA plasmids with sequences of bacterial or other origins, or the coding sequence or flanking sequences introduced have been modified from their native state in the crossable species where they are derived from.

Recent developments in the use of the CRISPR/Cas system in plant genome editing are reviewed by Wada et al., Metje-Sprink et al., Nadakuduti et al. [18, 29, 30]. A strategy that would inspire confidence in both consumers and regulators is one where the transformation method does not involve the use of DNA sequences at all. Thus, at least two DNA-free genome editing strategies have been reported. The first involves the use of viral RNA vectors. The second uses pre-assembled CRISPR/Cas, with only a short gRNA and no other nucleic acids.

An example of the first approach is presented by Ma et al. [31] who described an example of an RNA virus-vectored system. They engineered the negative-strand RNA virus, Sonchus yellow net rhabdovirus (SYNV) by inserting the CRISPR sequences for the guide RNA and Cas9 protein between the N and P genes of the virus. No selection marker was used. Infection was carried out by mechanical inoculation or by agroinfiltration of transformed *Agrobacterium* cells. Over 90% of plants regenerated from virus-infected tissue had the successful deletion of the target GFP gene used in the experiment [31]. The system must now be evaluated using an agriculturally important gene.

In the second approach, pre-assembled CRISPR/Cas9 ribonucleoproteins can be transfected into protoplasts or in vitro fertilized zygotes [32, 33]. This has been successfully done in rice zygotes [33]. However, the difficulty in the regeneration of whole plants from the protoplasts makes this method not applicable to many important species. The ribonucleoprotein or RNA may also be biolistically delivered into immature embryo cells or calli. This has been done with wheat [34, 35], maize [36] and rice [37]. However, the efficiency of editing is very low.

While there are thousands of CRISPR systems, most of the work has been done using the CRISPR/Cas9 system. However, the recently discovered system from *Prevotella* and *Francisella* (CRISPR/Cpf1, renamed CRISPR/Cas12) appears to be easier to adapt to DNA-free applications. This is mainly because the Cas 12a protein is smaller and will thus be easier to transfect into cells [29].

6. Conclusion

The availability of rapid and affordable sequencing technology together with tools for site-specific recombination are plants has made it possible for genetic engineers to design plant genetic engineering experiments with great precision. Cisgenic plants *sensus stricto* with agriculturally important traits will soon be a reality in the field. The first few varieties might have to go through the same regulatory processes as transgenic plants. Maybe the plants must go beyond the usual requirements and demonstrate their true equivalence to conventionally bred varieties. Beyond that, it is hoped that methodologies for cisgenesis will become standardized, and regulatory requirements of cisgenic plants will be equivalent to those of conventionally bred crops.

Acknowledgements

Financial support from CPUT is acknowledged.

Conflict of interest

No conflict of interest.

Author details

Richard Mundembe
Department of Biotechnology and Consumer Sciences, Cape Peninsula University of Technology, Cape Town, South Africa

*Address all correspondence to: mundember@cput.ac.za

References

[1] Singh OV, Ghai S, Paul D, Jain RK. Genetically modified crops: Success, safety assessment, and public concern. Applied Microbiology and Biotechnology. 2006;**71**:598-607. DOI: 10.1007/s00253-006-0449-8

[2] Wang GP, Yu XD, Sun YW, Jones HD, Xia LQ. Generation of marker- and/or backbone-free transgenic wheat plants via agrobacterium-mediated transformation. Frontiers in Plant Science. 2016;**7**:1324. DOI: 10.3389/fpls.2016.01324

[3] Keshavareddy G, Kumar ARV, Ramu VS. Methods of plant transformation: A review. International Journal of Current Microbiology and Applied Sciences. 2018;**7**(7):2656-2668. DOI: 10.20546/ijcmas.2018.707.312

[4] Rommens CM, Bougri O, Yan H, Humara JM, Owen J, Swords K, et al. Plant-derived transfer DNAs. Plant Physiology. 2005;**139**:1338-1349

[5] Schouten HJ, Krens FA, Jacobsen E. Cisgenic plants are similar to traditionally bred plants. EMBO Reports. 2006;**7**(8):750-753. DOI: 10.1038/sj.embor.7400769

[6] Bruening G, Lyons JM. The case of the FLAVR SAVR tomato. California Agriculture. 2000;**54**(4):6-7

[7] Mundembe R. Gene targeting and genetic transformation of plants. In: Sithole-Niang I, editor. Genetic Engineering. Rijeka: InTech; 2013. pp. 50-59. DOI: 10.5772/56335

[8] Hwang H-H, Yu M, Lai E-M. *Agrobacterium*-mediated transformation: Biology and applications. The Arabidopsis Book. 2017;**15**:e0186. DOI: 10.1199/tab. 0186

[9] Tian X, Zhou B. Strategies for site-specific recombination with high efficiency and precise spatiotemporal resolution. Journal of Biological Chemistry. 2021;**296**:100509. DOI: 10.1016/j.jbc2021.100509

[10] Olorunniji FJ, Rosser SJ, Stark WM. Site-specific recombinases: Molecular machines for the genetic revolution. The Biochemical Journal. 2016;**473**:673-684. DOI: 10.1042/BJ20151112

[11] Hardison RC. Recombination of DNA. State College, PA: The Pennsylvania State University; 2021

[12] Lieber MR. The mechanism of human nonhomologous DNA end joining. The Journal of Biological Chemistry. 2007;**283**(1):1-5. DOI: 10.1074/jbc.R700039200

[13] Ranjha L, Howard SM, Cejka P. Main steps in DNA double-strand break repair: An introduction to homologous recombination and related processes. Chromosoma. 2018;**127**(2):187-214. DOI: 10.1007/s00412-017-0658-1

[14] Chang HHY, Pannunzio NR, Adachi N, Lieber MR. Non-homologous DNA end joining and alternative pathways to double-strand break repair. Nature Reviews Molecular Cell Biology. 2017;**18**:495-506

[15] Jayaram M, Grainge I. Introduction to site-specific recombination. In: Mullany P, editor. The Dynamic Bacterial Genome (Advances in Molecular and Cellular Microbiology). Cambridge: Cambridge University Press; 2005. pp. 33-82. DOI: 10.1017/CBO9780511541544.002

[16] Ishino Y, Shinagawa H, Makino K, Amemura M, Nakata A. Nucleotide sequence of the *iap* gene responsible for alkaline phosphatase isozyme conversion in *Escherichia coli*, and identification of the gene product. Journal of Bacteriology. 1987;**169**(12):5429-5433

[17] Shmakov S, Smargon A, Scott D, Cox D, Pyzocha N, Yan W, et al. Diversity and evolution of class 2 CRISPR-Cas systems. Nature Reviews Microbiology. 2017;**15**:169-182. DOI: 10.1038/nrmicro.2016.184

[18] Wada N, Ueta R, Osakabe Y, Osakabe K. Precision genome editing in plants: State-of-the-art in CRISPR/Cas9-based genome engineering. BMC Plant Biology. 2020;**20**:234-246. DOI: 10.1186/s12870-020-02385-5

[19] Rommens CM, Humara JM, Ye J, Yan H, Richael C, Zhang L, et al. Crop improvement through modification of the plant's own genome. Plant Physiology. 2004;**135**:421-431

[20] Espinoza C, Schlechter R, Herrera D, Torres E, Serrano A, Medina C, et al. Ciscgenesis and intragenesis: New tools for improving crops. Biological Research. 2013; **46**:323-331

[21] Holme IB, Wendt T, Holm PB. Intragenesis and cisgenesis as alternatives to transgenic crop development. Plant Biotechnology Journal. 2013;**11**:395-407. DOI: 10.1111/pbi.12055

[22] Cardi T. Cisgenesis and genome editing: Combining concepts and efforts for a smarter use of genetic resources in crop breeding. Plant Breeding. 2016;**135**:139-147. DOI: 10.1111/pbr.12345

[23] Schaart JG, Krens FA, Pelgrom KT, Mendes O, Rouwendal GJ. Effective production of marker-free transgenic strawberry plants using inducible site-specific recombination and a bifunctional selectable marker gene. Plant Biotechnology Journal. 2004; **2**:233-240

[24] Joshi SG, Schaart JG, Groenwold R, Jacobsen E, Schouten HJ, Kren FA. Functional analysis and expression profiling of HcrVf1 and HcrVf2 for development of scab resistant cisgenic and intragenic apples. Plant Molecular Biology. 2011;**75**:579-591

[25] Vanblaere T, Szankowski I, Schaart J, Schouten HJ, Flachowsky H, Broggini GA, et al. The development of a cisgenic apple plant. Journal of Biotechnology. 2011;**154**:304-311

[26] Cody JP, Graham ND, Zhao C, Swyers NC, Birchler JA. Site-specific recombinase genome engineering toolkit in maize. Plant Direct. 2020;**4**:1-9. DOI: 10.1002/pld3.209

[27] Gadaleta A, Giancaspro A, Blechl AE, Blanco A. A transgenic durum wheat line that is free of marker genes and expresses IDY10. Journal of Cereal Science. 2008;**48**:439-445

[28] Holme IB, Dionisio G, Brinch-Pedersen H, Wendt T, Madsen CK, Vincze E, et al. Cisgenic barley with improved phytase activity. Plant Biotechnology Journal. 2012; **10**:237-247

[29] Metje-Sprink J, Menz J, Modrxejewski D, Sprink T. DNA-free genome editing: Past, present and future. Frontiers in Plant Science. 2019;**9**:1957. DOI: 10.3389/fpls.2018.01957

[30] Nadakuduti SS, Enciso-Rodriguez F. Advances in genome editing with CRISPR systems and transformation technologies for plant DNA manipulation. Frontiers in Plant Science. 2021;**11**:637159. DOI: 10.3389/fpls.2020.637159

[31] Ma X, Zhang X, Liu H, Li Z. Highly efficient DNA-free plant genome editing using virally delivered CRISPR-Cas9. Nature Plants. 2020;**6**(7):773-779. DOI: 10.1038/s41477-20-0704-5

[32] Woo JW, Kim J, Kwon SI, Corvalán C, Cho SW, Kim H, et al.

DNA-free genome editing in plants with preassembledCRISPR-Cas9 ribonucleoproteins. Nature Biotechnology. 2015;**33**:1162-1164. DOI: 10.1038/nbt.3389

[33] Toda E, Koiso N, Takebayashi A, Ichikawa M, Kiba T, Osakabe K, et al. An efficient DNA- and selectable-marker-free genome-editing system using zygotes in rice. Nature Plants. 2019;**5**:363-368. DOI: 10.1038/s41477-019-0386-z

[34] Zhang Y, Liang Z, Zong Y, Wang Y, Liu J, Chen K, et al. Efficient and transgene-free genome editing in wheat through transient expression of CRISPR-Cas9 DNA or RNA. Nature Communications. 2016;**7**:12617. DOI: 10.1038/ncomms12617

[35] Liang Z, Chen K, Li T, Zhang Y, Wang Y, Zhao Q, et al. Efficient DNA-free genome editing of bread wheat using CRISPR-Cas9 ribonucleoprotein complexes. Nature Communications. 2017;**8**:14261. DOI: 10.1038/ncomms14261

[36] Svitashev S, Schwartz C, Lenderts B, Young JK, Mark CA. Genome-editing in maize directed by CRISPR-Cas9 ribonucleoprotein complexes. Nature Communications. 2016;**7**:13274. DOI: 10.1038/ncomms13274

[37] Li S, Li J, He Y, Xu M, Zhang J, Du W, et al. Precise gene replacement in rice by RNA transcript-templated homologous recombination. Nature Biotechnology. 2019;**37**:445-450. DOI: 10.1038/s41587-019-0065-7

Chapter 5

Genetic Conservation and Importance of Ginger in Ethiopia

Girma Hailemichael, Mohammedsani Zakir and Melaku Addisu

Abstract

Ethiopia has conducive agro-ecology and is capable to produce various spices, including ginger. Spices research in Ethiopia with various limitations did a lot on ginger technology development and achieved significant results. Acquiring significant number of ginger accessions from domestic and foreign source, variety development, pre and postharvest management practices including protection parts are available. Ethiopians have high spices consumption culture and significant volume of ginger used locally and this can be considered as one of the opportunities. In addition to this, there is significant export of ginger and generating 22.6 million USD in 2008. Though there is such potential for production and export earnings, still there are lots of challenges that keep the performance of the commodity very low. The local grouping of the ginger materials, and the selection and release of two ginger cultivars from Jimma Agricultural Research center/Tepi Agricultural Research Sub Center, confirmed that the country has a high diversity of germplasm. This chapter reviews the ginger germplasm enhancement, conservation, variety development, status of ginger breeding, diversity of ginger, ginger tissue culture, ginger biotechnology.

Keywords: ginger, ginger breeding, varieties, diversity, invitro, Ethiopia

1. Introduction

Ginger (*Zingiber officinale* Rosc) is a monocotyledonous, herbaceous, and tropical plant belonging to the family, Zingiberaceae. It has a long and well-documented history of both culinary and medicinal use throughout the world history, especially in medical care in China, India, and Japan. Ginger is a self-incompatible plant that is characterized by high rates of infertility; thus, its genetic diversity occurs through processes of mutation and natural selection alone. Since ginger is one of the major spices, knowing the diversity in depth can contribute to conservation and its use in breeding programs [1]. In Ethiopia, ginger has been used as a fragrance, carminative, and stimulant and has become a major spice in both the local and export markets. Arabs brought ginger from India to East Africa in the thirteenth century [2], and ginger has since been known in Ethiopia and is cultivated primarily in the humid regions of the Southern Nations (SNNPRS). To a lesser extent, ginger production has expanded to Western Oromia and in Amhara region chilga areas. Commercial production of ginger by smallholder farmers is practiced in South region Kambata-Tambaro, Wolaita, and Hadiya zones. According to statistics from the Ministry of Agriculture and Rural Development, 99% of crop production occurred in the Southern Nations, Nationalities, and People's States [3, 4].

Ginger prefers a warm and humid climate, and most soils have sufficient water retention and ventilation. It is cultivated in the tropics up to 1500 m above sea level, but partial shade also increases its yield. The base temperature requirement for ginger is 13°C and the upper limit is 32°C/27°C (day/night), whereas the favorable range is 19–28°C. The optimum soil temperature for germination is between 25 and 26°C, and 27.5°C, which is required for growth. Ginger research has been done in Ethiopia for many years. Jimma Agricultural Research Center (JARC) under its sub-center (before), Tepi Agricultural Research Sub-center was responsible for improving ginger's genetic resources and developing varieties. In this responsibility and action, JARC officially launched two ginger varieties called Yali and Boziab in 2007. The varieties give high yield (200–250 Q/ha of fresh rhizome). Essential oil (1.8–2.5% v/w) and oleoresin (6.01–8.22% w/w) content from these varieties satisfy quality standards. Various technologies on agronomic production and seed rhizome and dried yield rhizome were generated.

Three types of the ginger product known as fresh rhizome, dried rhizome, and extracted rhizome are supplied to the market. Ethiopia used to export fresh ginger to Egypt, Saudi Arabia, and Yemen. Fresh Ethiopian ginger has been reported to be an excellent product with good color and quality and long shelf life [2]. Dried ginger is Ethiopia's most popular ginger product and is mass exchanged by all market participants at various stages of marketing, from local assembly to the export market. The third type of ginger product on the market is extracted ginger products which include powdered ginger, essential oils, oleoresin, etc. [3].

Ginger production and productivity in Ethiopia was challenged by a number of factors; shortage of varieties with full production packages, postharvest practices, lack of value addition, and poor/limited large-scale investments. Demand for ginger increasing in the domestic and foreign market. Ethiopia's diverse climate and soil types greatly contribute to ginger production and these all are good opportunities to conserve ginger genetic resources, improve production and productivity and bring significant income and hard currency for the country.

2. Breeding and genetics

2.1 Germplasm enhancement

The first step in doing good breeding research is to obtain a sufficient number of accessions in the relevant or target crop. To achieve this goal in the case of ginger, two key tactics were used: collecting from domestic sources and introduction from abroad. The spices research team started before four and half decades since the start of coffee research as diversification in Jimma Agricultural Research Center (JARC). A few years later, the introduction of ginger accessions began, and at the same time, collecting from several possible ginger-growing sites across the country was undertaken. The collection is always conducted in collaboration with the expertise from Institute of Biodiversity and Conservation (IBC). The significant number of ginger accessions was attained from this activity except that it has been challenged by bacterial wilt of ginger.

2.2 Conservation

Collected and introduced accessions of ginger were maintained in research plots of Jimma Research Center and mainly in Tepi Agricultural Research Sub-Center before, currently upgraded to Tepi Agricultural Research Center. Despite the threat of the ginger bacterial wilt disease, which destroyed more of the collections, more than 90 accessions have been gathered and some introduced since the research began. Ginger germplasms that have been collected and introduced are evaluated using a variety of criteria. The table below (**Table 1**) displays a sample of the accessions and their sources. We recently established 45 collections in the JARC research plot, which are being used for various research purposes related to ginger bacterial wilt management.

2.3 Variety development

Rigorous adaptation test of ginger accessions has been conducted in different agro-ecologies, suitable propagation parts, and status to sprouting and subsequent field performance and rhizome yield. Pest reactions and quality were some of the traits used in the evaluation. According to the results obtained [5, 6], the varieties were proved promising and widely employed by users and distributed throughout the potential producing areas. Further evaluation continued to achieve better varieties. Ginger bacteria diseases have been devastating to the materials. This

No.	Entry code	Origin	No.	Entry code	Origin
1	Ging.28/79	Mauritius	24	Ging.30/86	Collected
2	Ging.36/79	Australia	25	Ging.24/86	Collected
3	Ging.41/79	Riodjenero	26	Ging.85/86	Collected
4	Ging.316/73	Surinam	27	Ging.45/86	Collected
5	Ging.296/79	Rafinufa	28	Ging.75/00	Collected
6	Ging.305/73	Collected	29	Ging.61/00	Collected
7	Ging.25/86	Collected	30	Ging.307/72	Collected
8	Ging.28/86	Collected	31	Ging.087/00	Collected
9	Ging.61/86	Collected	32	Ging.15/79	Rafinufa
10	Ging.10/86	Collected	33	Ging.38/79	Australia
11	Ging.48/86	Collected	34	Ging.39/79	Australia
12	Ging.57/86	Collected	35	Ging.180/73	Collected
13	Ging.84/86	Collected	36	Ging.181/73	Collected
14	Ging.70/00	Collected	37	Ging.47/86	Collected
15	Ging.74/00	Collected	38	Ging.53/86	Collected
16	Ging.41/00	Collected	39	Ging.58/86	Collected
17	Ging.16/79	Rafinufa	40	Ging.59/86	Collected
18	Ging.37/79	Australia	41	Ging.56/86	Collected
19	Ging.40/79	Riodjenero	42	Ging.54/86	Collected
20	Ging.141/73	Australia	43	Ging.26/86	Collected
21	Ging.190/73	Collected	44	Ging.86/00	Collected
22	Ging.29/86	Collected	45	Ging.63/00	Collected
23	Ging.52/86	Collected			

Sources: [4, 5].

Table 1.
Germplasms and varieties of ginger in Ethiopia.

time, more than 45 accessions of ginger are being maintained for future disease management research [7].

2.4 Status of ginger breeding

In Ethiopia, ginger breeding was started as a part of coffee diversification in JARC, since the inception of coffee research in 1969 [8]. Some preliminary research on local and introduced ginger germplasms indicated the existence of genetic variability in their morphological traits, rhizome yield, oil and oleoresin contents [9]. High variability was observed among ginger cultivars and/or accessions for plant height, rhizome yield, oil content, and oleoresin. According to Momina *et al.* [8] there was high genetic diversity in local ginger germplasms.

2.5 Diversity of ginger in Ethiopia

The introduction of ginger to Ethiopia for a long time (thirteenth century), made the country to have diverse genetic resources. Variability (in morphology and quality) was reported by Momina *et al.* [8]. Such status of diverse genetic resources is crucial for breeding purpose. Southern Nation and Nationalities regional state are often understood to be major areas of ginger germplasms. Indicating that there is high diversity of ginger in Southern Nation and Nationalities regional state Wolaita zone farmers group local varieties into Masculine and Feminine [3]. Also, farmers in Kambata-Tambaro recognized one local genotype called Hargema (**Figure 1**). There is some similarity among these materials.

Production of local ginger materials in Wolaita and Hadiya zones has been since time immemorial and various local landraces were common in different areas (Bilbo and Volvo (**Figure 2**) introduced to the area recently (in 1998). According to Geta and Kifle [3] seed transfer and distribution as informal ways (farmer-to-farmer) remained very common.

The local grouping of the ginger materials discussed here (**Table 2**), and the selection and release of two ginger cultivars from Jimma Agricultural Research center/Tepi Agricultural Research Sub Center, confirmed that the country has a high diversity of the germplasm. This needs further research to exploit ginger genotypes with different quality parameters and special traits.

2.6 In vitro propagation for maintenance of ginger

Propagation of two ginger cultivars by tissue culture is one of the strategy to improve production and productivity by overcoming ginger bacterial wilt. The study was carried out with the objective of assessing the potential of axillary buds and shoot tips as explant sources and determination of suitable growth regulators for in vitro propagation. MS medium with four levels of benzyl adenine (BA) and kinetin was used for shoot multiplication in combination with two explant sources. Shoot tip explants on 2 mg l^{-1} BA and 1.2 mg l^{-1} kinetin was found to be better than other explant-media combinations. Consecutively, the plantlets developed an average of 8.75 roots within 4 weeks of the culture period and performed well in greenhouse acclimatization and field operations. In vitro propagation of the Yali and Boziab was proved possible with this explant and media combinations. Parameters such as number of leaves and dry weight of plantlets regardless of the varietal difference in comparison to axillary bud was higher. Similarly, shoots cultured on MS medium with 1 mg l^{-1} NAA alone developed vigorous roots. Plantlets produced by this propagation protocol were successfully acclimatized within 4 weeks of hardening. The acclimatization procedure has been supported with the application of

Figure 1.
Feminine Wolaita (left), Masculine Wolaita (right). Source: [3]

Figure 2.
Subterranean parts of Bilbo with two taproots (left), Volvo with a single taproot (right). Source: [10].

shade nets (at 30 and 70% shade level) and polythene under the greenhouse condition. Subsequently, the seedlings have survived well under field conditions [12].

2.6.1 Ginger tissue culture

In Ethiopia, infection with *Ralstonia solanacearum* has resulted in significant losses of ginger rhizomes. To have successful ginger cultivation, disease-free planting material generation is required and practiced. Plant tissue culture technology has proven to be effective in the commercial production of pathogen-free plants as well as the preservation of rare and endangered species' genetics (**Table 3**). The initial surface sterilization experiment was effective when 0.7–1.5 cm shoot tips were treated with 70% ethanol for 5 min followed by double sterilization with 5% active chlorine concentration of local bleach (Clorox), for 15 min under aseptic condition. This treatment cleans bacterial contamination more than 95%. Also, antibiotics

Category	Vernacular	Unique characteristics	Common characteristics
Local	Feminine "Wolaita"	• Highly palmated rhizomes • Large-sized rhizomes • More productive	• Drought tolerant • Less fertilizer requirement • Long postharvest storability (up to 10 years) • Suitable for perennial harvesting • Less susceptible to mold development when subject to sun-drying at wet weather condition • More preferred to dry planting
	Masculine "Wolaita"	• High fiber content • Highly pungent	• High harvesting cost • Less productive/unit area • Large number of prominent roots • High cost of root trimming during rhizome drying
	"Hargema" (Kambatigna)	• Highly pungent • Large number of prominent roots ('Amesalgier')	• Late maturing • Less preferred for fresh rhizome market • Low dry matter content (high degree of shrinkage upon drying) • Fast rate of drying • Less attractive appearance • Much weight loss with prolonged storage period

Category	Vernacular	Unique characteristics	Common characteristics
Introduced	"Bilbo" (Wolaitigna and Kambatiga)	One prominent root/digitally palmated rhizome	• Drought tolerant • Soil exhaustive • Short postharvest storability • Less suitable for perenniated harvesting, dries up with extended dry season as a result of less number of prominent roots • Highly susceptible to mold development when subject to sun-drying at wet weather condition • Less preferred to dry planting • Easy for manual harvesting; low harvesting cost • More productive/unit area • Low number of prominent roots • Low cost of root trimming during rhizome drying
	"VolVo" (Wolaitigna and Kambatigna)	• 2–3 prominent/tap roots/digitally palmated rhizomes • Relatively highly pungent	• Early maturing • Highly preferred at fresh rhizome market • High dry matter content (less degree of shrinkage upon drying) • Slow rate of drying • Highly attractive appearance • High demand at the central market • Easily peeling off of the skin at frequent overturning during the process of drying, which in turn, increases its susceptibility to mold development • Minimum weight loss upon prolonged storage period

Source: [10].

Table 2.
Characteristics of different ginger vernaculars cultivated in SNNPRS.

Antibiotics	Antibiotics concentration (mg/l)	Shoot number mean ± SD
Control	00 mg/l	8.00 ± 2.71abc
Ampicillin	130 mg/l	11.51 ± 1.29a
Ampicillin	160 mg/l	8.75 ± 1.70abc
Ampicillin	200 mg/l	6.74 ± 1.70bc
Ampicillin	250 mg/l	7.00 ± 1.40bc
Gentamaycine	130 mg/l	6.00 ± 1.40c
Gentamaycine	160 mg/l	5.70 ± 1.50c
Gentamaycine	200 mg/l	5.75 ± 1.70c
Gentamaycine	250 mg/l	6.75 ± 0.50bc
Streptomycin	130 mg/l	6.75 ± 1.25bc
Streptomycin	160 mg/l	7.00 ± 0.00bc
Streptomycin	200 mg/l	6.60 ± 0.57bc
Streptomycin	250 mg/l	6.50 ± 0.57bc
Tetracycline	130 mg/l	8.00 ± 1.5bc
Tetracycline	160 mg/l	7.70 ± 0.81bc
Tetracycline	200 mg/l	7.20 ± 0.50bc
Tetracycline	250 mg/l	6.00 ± 1.15c

Numbers are mean and SD of four replicates (four plant in each culture jars).
Means followed by the same letter within a column are not significantly different by Tukey's test at α = 0.05%.
Source: [11].

Table 3.
Antibiotics treatment of plant material results in shoots free from R. solanacearum.

(tetracycline and streptomycin) reduced the contaminants thereby increase the survival rate of the plantlets.

2.6.2 Ginger biotechnology

Consistent and increasing demand for clean planting material from improved cultivars of ginger is persistent. Providing the required through indigenous techniques of propagation is incompetent owing to inefficient production and transmission of disease. In this regard, to evaluate the potential of shoot tips and axillary buds and to determine the appropriate growth regulators for propagation In vitro was attempted in two ginger cultivars [12]. From the study, it is reported that the better shoot multiplication of average for each explant was 7 shoots after culturing for 6 weeks attained on BA (2 mg/l) and kinetin (1 mg/l) with a huge significant difference in observation between explant source and growth regulator used. Successful root induction (8.75) in 4 weeks of culture and well acclimatization and field survival were noticed in the plantlets generated. Berihu [13] reported on disinfection of ginger sprout buds and disease screening with tests that have resulted in disease-free plantlets of ginger through mass propagation and commercialization to customers. A Series of washing steps with $CuSO_4$ with Tween 20 with different time intervals and flashing with sterile water has resulted in effective disinfection. Biochemical examination and serological test via NCM-ELISA for cleaning of disease in vitro and mass propagation of ginger for samples tried and yielded successful raising ginger sample. Another study on In vitro micropropagation of shoot tip explants by Selam *et al.* [14] using Ethiopian ginger cultivar to overcome the problem of unclean and unhealthy ginger planting material is published late. The wilt disease due to *Ralstonia solanacearum* Biovar 3 Race 4 has resulted in obtaining masses of profuse planting materials free from disease. The study aims at revealing the effect of three sterilization agents namely RBK (0.25% w/v), NaOCl (0.50% v/v) and ethanol (70% v/v) in mixture with $HgCl_2$ (0.25%). Study of efficacy for 4 antibiotics (broad-spectrum) in combination to control contaminants of bacteria with shoot tip explants of ginger and the effect of antibiotics performance on the shooting of explants of cultivar have been studied. Live explants (70%) with 80% free from contamination were obtained after 3 weeks of incubation from 20 min exposure to 0.50% v/v NaOCl continued by 0.25% $HgCl_2$. Of all the combinations tried the highest (7.10 ± 0.36 and 7.51 ± 0.27, respectively) mean micro shoots per explant and mean length of shoot (4.2 and 3.56 cm) were obtained at cefotaxime (50 mg/l) and cefotaxime with streptomycin (25 mg/l). The results presented in this study could provide some basic foundation for optimizing protocols in sterilizing explants and can effectively control the bacterial contaminants in ginger cultivar for large-scale micropropagation [15].

3. Conclusion

There is attractive local and foreign market for different types of ginger products; dry, sliced, extracted oleoresin, and essential oil content. Government, research, NGOs, and private sectors need to work on the conservation and maintenance of the ginger germplasms in Ethiopia.

Acknowledgements

The authors would like to acknowledge all the parties working on spices/ginger and contributed for writing of this article.

Author details

Girma Hailemichael*, Mohammedsani Zakir and Melaku Addisu
Jimma Agricultural Research Center, Jimma, Ethiopia

*Address all correspondence to: ghailemichael1973@gmail.com

References

[1] Cardoso ES, Rossi AAB, Pedri ECM, Zortéa KEM, Tiago AV, Rocha VD, et al. Genetic diversity of *Zingiber officinale* (Zingiberaceae) germplasm grown in urban and rural backyards in Mato Grosso, Brazil. Genetics and Molecular Research. 2020;**19**(2):1-12

[2] Jansen PCM. Spices, Condiments and Medicinal Plants in Ethiopia, Their Taxonomy and Agricultural Significance. Wageningen: PUDOC; 1981. pp. 1-132

[3] Geta E, Kifle A. Production, processing and marketing of ginger in Southern Ethiopia. Journal of Horticulture and Forestry. 2011; **3**(7):207-213

[4] Girma H, Digafe T, Edossa E, Belay Y, Weyessa G. Spices Research Achievements. Revised Edition. Addis Abeba Ethiopia: EIAR; 2008. pp. 12-23

[5] Edossa E. Research Achievements and Experiences. Research Report No. 3. Addis Abeba, Ethiopia: Ethiopian Institute of Agricultural Research (EIAR); 1998. pp. 6-14

[6] Girma H, Habtewold K, Haimanot M. Spices research achievements, challenges and future prospects in Ethiopia. Academic Research Journal of Agricultural Science and Research. 2016;**4**(1):9-17

[7] Zakir M, Girma H, Mesfin S, Melaku A. Status of ginger (*Zingeber officinale*) research and production challenges and future prospects in Ethiopia; a review. Academic Research Journal of Agricultural Science and Research. 2018;**6**(5):276-290

[8] Momina A, Sintayehu A, Girma H, Abush T. Variability of ginger (*Zingiber officinale Rosc.*) accessions for morphological and some quality traits in Ethiopia. International Journal of Agricultural Science. 2011;**6**(6):444-457

[9] Hailemichael G, Tilahun D. Current Status of Spice Research. Addis Ababa, Ethiopia: Ethiopian Agricultural Research Organization, Jimma Agricultural Research Center; Tepi Agricultural Research Sub-Center; 2004

[10] Teferra TF. Ginger (*Zingiber oficinale* Rosec.) Production, Postharvest Handling, Processing and Marketing: *A Comprehensive Extension Package Manual*. Addis Ababa, Ethiopia: Farm Africa; 2015

[11] Markos T. Effect of antibiotics in eliminating bacterial wilt (*Ralstonia solanacearum*) from in vitro propagated ginger (*Zingiber officinale* Roscoe) [thesis submitted in partial fulfillment of the requirements for the degree of Master of Science in Biotechnology]. Addis Ababa, Ethiopia: School of Graduate Studies of Addis Ababa University; 2019

[12] Biruk A, Wondyifraw T, Kassahun B. In vitro propagation of Ethiopian ginger (*Zingiber officinale* Rosc.) cultivars: Evaluation of explant types and hormone combinations. African Journal of Biotechnology. 2012;**11**(16):3911-3918. DOI: 10.5897/AJB10.1962

[13] Berihu M. In vitro diseases cleaning, micro-propagation and field establishment of ginger (*Zingiber officinale* Rosc.) cultivars. International Journal of Pharmaceutical Science and Research. 2018;**3**(4):8-11

[14] Selam T, Subban P, Zenebe T, Desta BS. Assessing the efficacy of broad-spectrum antibiotics in controlling bacterial contamination in the In vitro Micropropagation of Ginger (*Zingiber officinale* Rosc). The Scientific

World Journal. 2020;**8**:6431301. DOI: 10.1155/2020/643 1301

[15] Paramesh H, Abreham CA. Plant Biotechnology in Ethiopia: Current Status, Opportunities and Challenges. Plant Cell Biotechnology and Molecular Biology. 2021;**22**(13&14):136-156

Chapter 6

Crop Biotechnology and Smallholder Farmers in Africa

Endale Gebre Kedisso, Nicolas Barro, Lilian Chimphepo, Tahani Elagib, Rose Gidado, Ruth Mbabazi, Bernard Oloo and Karim Maredia

Abstract

The tools of genetic engineering and modern biotechnology offer great potential to enhance agricultural productivity, food and nutritional security, and livelihoods of millions of smallholder farmers in Africa. Large and long-term investments have been made in several countries in Africa to access, develop, and commercialize safe biotechnology crops derived through modern biotechnology. This chapter presents case studies of biotechnology applications and progresses achieved in six countries in Sub-Saharan Africa including Burkina Faso, Ethiopia, Kenya, Malawi, Nigeria, Sudan, and Uganda targeting to address biotic and abiotic constraints faced by smallholder farmers and malnutrition. Based on the past 20 years of experience, the chapter identifies constraints, challenges, and opportunities for taking safe biotechnology crops to smallholder farmers in Africa.

Keywords: biotechnology, biosafety, genetic engineering, GMOs, Burkina Faso, Ethiopia, Kenya, Malawi, Nigeria, Sudan, Uganda

1. Introduction

1.1 Smallholders' agricultural production and productivity in Africa

In Africa, smallholder agriculture is predominant and agricultural growth and poverty reduction are subjects closely associated with growth in smallholder agriculture for some time to come. An estimated 41 million smallholders [1] are the major source of food for nearly all rural and most urban dwellers in Africa. In Sub-Saharan Africa (SSA), most smallholders own less than two hectares holding of cultivable land and are challenged by the low productivity and production constraints in the middle of the unprecedented rising need for more food, feed, and raw material for industry. The SSA region alone has a quarter of the world's arable land endowment but produces only 10% of world agricultural output [2]. Unlike smallholders in Asia who dominantly grow few crops such as rice and wheat, African farmers experience diverse farming systems and grow very diverse crops that include maize (*Zea mays*), sorghum (*Sorghum* sp) millet (*Penisetum* sp), wheat (*Triticum aestivum*), and rice (*Oryza sativa*); pulses such as soybean (*Glycine max*), cowpea (*Vigna unguiculata*), beans (*Phaseolus* sp.), groundnut (*Arachis hypogaea*), and other crops such as cassava (*Matnihot esculentus*), sweet potato (*Ipomoea*

Farming systems	% of region		Principal livelihoods*
	Land area	Agric. population	
Irrigated	1	2	Rice, cotton, vegetables, rain-fed crops, cattle, poultry
Tree Crop	3	6	Cocoa, coffee, oil palm, rubber, yams, maize
Forest-Based	11	7	Cassava, maize, beans, cocoyams
Rice-Tree Crop	1	2	Rice, banana, coffee, maize, cassava, legumes, livestock, off-farm work
Highland Perennial	1	8	Banana, plantain, enset, coffee, cassava, sweet potato, beans, cereals
Highland Temperate Mixed	2	7	Wheat barley, tef, peas, lentils, broad beans, rape, potatoes,
Root Crop	11	11	Yams, cassava, legumes, off-farm work
Cereal-Root Crop Mixed	13	16	Maize, sorghum, millet, cassava, yams, legumes, cattle
Maize Mixed	10	15	Maize, tobacco, cotton, cattle, goats, poultry,
Agro-Pastoral Millet/Sorghum	8	8	Sorghum, pearl millet, pulses. Sesame and livestock
Sparse (Arid)	17	1	Irrigated maize, vegetables, date palms, cattle

**Source: FAO and World Bank, Rome and Washington DC 2006. (Adapted to show more crop-based farming system).*

Table 1.
Major farming systems of sub-Saharan Africa.

batatas), potato, (*Solanum tuberosum*), yam (*Dioscorea* sp), banana (*Musa* sp), cotton (*Gossypum* sp), and sugarcane (*Saccahrum officinarum*) (**Table 1**) [3].

Crop productivity in Africa specifically in the SSA region is below the world average (**Figure 1**) and the region constitutes the highest number of food-insecure population (35.5% of its population) of whom 21.3% are severely insecure [4] rendering the region increasingly dependent on imported food. Due to this and

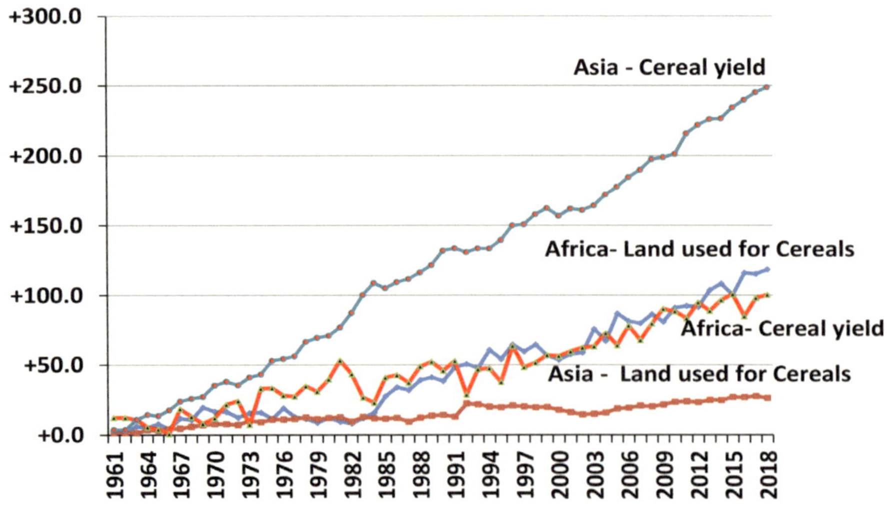

Figure 1.
Change (percent increase) of cereal yield and land used for cereal production. (Data source: Computed from Food and Agriculture Organization (FAO) of the United Nations. 2019 Report).

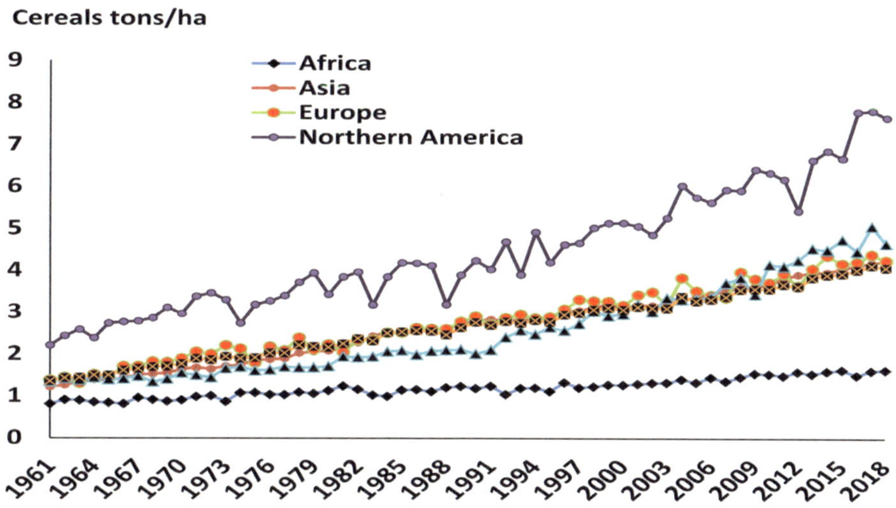

Figure 2.
Yield (t/ha) trends of cereal production in different regions of the world. (Data Source: Food and Agriculture Organization of the United Nations. 2019 Report).

other factors about 39 countries of the SSA account for the largest number of food-insecure people: 424.5 million (40.5% of the region's population) in the year 2020 [5]. It can also be seen that during the period 1961–2018, cereal yield in Africa has grown only one fold compared to a 2.5 fold increase in Asia, which had only 26.3% area increase compared to Africa with 1.2 fold increase (**Figure 1**). Therefore, whatever growth there has been in cereal production in Africa, it was largely due to land expansion in contrast to Asia. Food insecurity is forecasted to worsen due to climate change impacts and recurrent drought unless proper and quick measures are implemented [6]. The region will have a shortfall of nearly 90 million metric tons of cereals by the year 2025 if current agricultural practices remain unchanged. Productivity trends do not promise a better future for cereals and roots and tuber crops as can be seen from cereal performance during the period 1961–2018 average yield based on FAOSTAT data 2020 (**Figure 2**).

However, more factors are known to involve in constraining smallholder farmers' crops production and cause yield gaps. Low crop productivity is often related to biotic stresses such as those caused by insect pests, diseases, and weeds as well as the inherent low-yielding potential of varieties, and abiotic stresses caused by soil-related and climatic problems such as moisture stress and drought. The latter is a pronounced problem of vast marginal and drier agriculture areas of SSA. Crops grown in such marginal environments are exposed to frequent severe growing conditions. Each factor is responsible for substantial yield losses annually by smallholder farming. Furthermore, yield gains associated with high-yielding varieties if found much lower in SSA partly due to inadequate inputs, poor infrastructure, and market outlet including weak extension services. Thus, poor availability of improved technology packages (improved seeds, irrigation, fertilizers, and pesticides) makes it hard for millions of smallholder farmers to produce surplus and escape the subsistence type of life.

Successful mitigation of these biotic and abiotic constraints and institutional limitations affecting agricultural growth is a task that not only requires political will and sustained commitment by country governments in Africa, but also a stronger global collaborative effort to realize enhanced applications of modern technologies to complement and transform the conventional interventions efforts underway. Increased investments in agricultural R&D and fast-tracking the use of innovative technologies such as conventional as well as modern biotechnology and proven useful readily available biotechnology products is extremely needed to solve smallholder farmers' crop productivity problems. As such agricultural biotechnology offers enormous opportunities through innovative ideas, techniques, and processes to drive innovative solutions highly relevant for the needs of smallholder farmers in Africa [7]. Medium to long-term benefits of using advanced techniques of biotechnology that include tissue culture, micropropagation, gene, and marker discovery, genomics, genetic engineering, genome-editing, bioinformatics, and others through enhancing crop breeding including indigenous crop species cannot be overemphasized [8]. This chapter focuses on the deployment of modern biotechnology such as genetic engineering tools and products as well as challenges facing adopting countries in developing Africa. It also presents case studies of agricultural biotechnology uses and progresses in six countries in SSA focusing on the use of safe biotechnology crops to solve key biotic and abiotic constraints faced by smallholder farmers in the respective countries.

1.2 Promises of biotechnology to smallholder farmers

The rapid advancements in the field of biotechnology offer promising alternatives to the approaches of crop improvement. Biotechnology complements and makes the conventional breeding efforts in crops efficient through precise identification and introgression of genes in a much shorter time period. The integration and development of biotechnology research in national research programs is now a prerequisite for current and most of the future science-based sustainable genetic improvement of crops for various purposes including, food and nutritional security, improving post-harvest and industrial qualities of cereals, horticultural and forage crops.

It is clear that smallholder farmers in African countries are currently not benefiting enough from modern biotechnology, which can be applied to transform their crop production and productivity and bring about livelihood improvements. Most national research programs in Africa have not yet acquired research and regulatory capacity and skills to integrate advanced science and cutting-edge technologies in their research portfolio to solve farmers' production problems. Although progress is registered in biotechnology capacity building in some countries, it is far from adequate. Governments' investment in agricultural research and development is generally low [9]. Crop productivity problems under smallholder farmers' conditions are often caused by low-level use of improved technologies and damage to crops caused by biotic and abiotic stresses as described earlier. The biotic and abiotic stresses challenging crop productivity are being tackled by biotechnology globally and several crop varieties with novel traits have been successfully developed and commercialized in more than 25 countries around the world to solve particular production problems of farmers.

1.3 Crop improvement programs in Africa

Food security and prosperity in Africa depend much on its agricultural performance. Ensuring sustainable development in agriculture is critically dependent on a sustainable technology supply and uptake. Despite the strong need for robust agricultural research, capable of tackling production constraints under challenging agricultural environments, African countries have not shown much progress in their national research capabilities to respond to food security issues and meet the overarching national strategic goals for sustainable development [9]. Strategic measures pursued to realize latecomer advantages in using modern biotechnology to enhance crop improvement and exploiting existing commercialized novel biotechnology products proven safe and impactful, is weak.

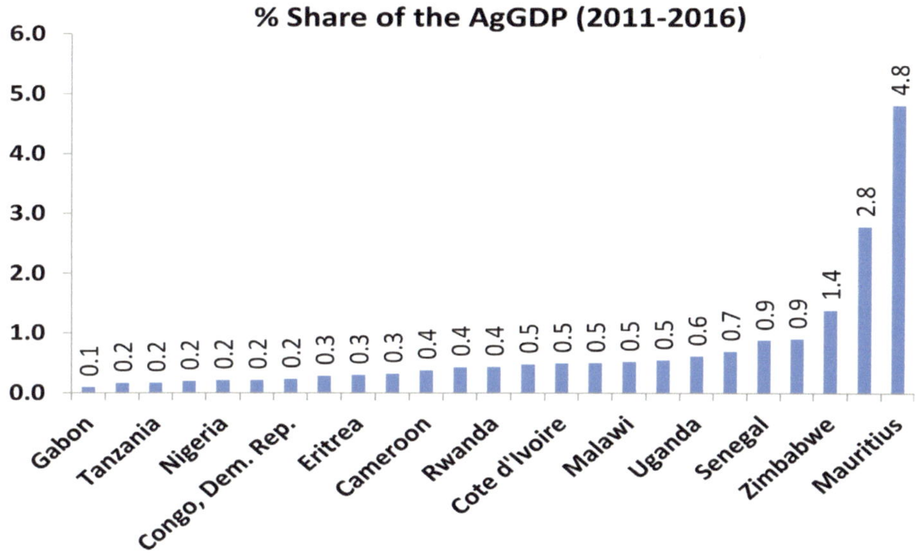

Figure 3.
Some SSA countries and their R&D investment share as a percent of AgGDP (except the top ranking the last three countries, all the others are selected only for representation of the rest). Source: Data sourced from ASTI [10].

Reports show declining government R&D spending in the agricultural sector recently from 0.59% in 2000 to 0.39% in 2016 in the SSA [10]. Thirty-three of the 44 SSA countries have less than the minimum investment target of 1% AgGDP (**Figure 3**) recommended by the African Union and United Nations [11]. Thus, most national programs in Africa were not able to maintain up-to-date capacity in trained human resources and facilities to translate scientific research into useful products impacting agricultural growth. Conventional crop improvement programs are increasingly requiring support from biotechnology to effectively respond to changing market demands. Therefore, African government should play a key role to strengthen national programs and maintain strong regional and global collaborative partnerships and expedite knowledge and technology transfer. Allowing more regional integration can help to ensure smoother collaboration, transfer of suitable technologies, data and information, and allows improved access to products at an affordable price and quality [12].

Most African countries have not created the necessary incentives for high-end modern biotechnologies to get well integrated in the research and development profile of national programs and create opportunities for new products to get to market. Instead, they depend on other countries that have decided to invest and strengthen their R&D. They are not taking advantage of this to enable national programs to expedite adoption and use of better and diverse technologies through quick testing and approval processes. Biotech products are rapidly expanding to include not only farmers' interest but getting more diversified targeting the interest of industry and consumers [13]. Therefore, a further declining trend of investment in agricultural R&D over the past 15–20 years in the developing countries with few countries in exception is alarming [14]. In countries with advanced economies where public financial outlay for R&D has lagged, the private sector has been investing heavily in genomic sciences and techniques that enable faster and more efficient delivery of improved crops to farmers, the value chain, and consumers, targeting business opportunities and crops with the greatest returns to investment [7]. However, many 'orphan' or underutilized indigenous crops in developing countries have been forgotten and their diversity is threatened [7]. It is highly challenging to rectify this imbalance between public and private research investment and ensure that crops including indigenous species are improved and conserved thus equally benefiting from modern biotechnology.

Against all odds and considerable skepticism in African countries even after three decades of the phenomenal growth of modern biotechnology and wider adoption of safe biotechnology crops globally, some countries have moved forward and strengthened capacity in biotechnology and related fields of biosafety, food safety, and intellectual property (IP) management to reap the benefits of integrating the advanced sciences. The recent progress in approvals of several biotechnology crops in Africa can reverse the delay in the near future [15–18].

2. Role of agricultural biotechnology: narrowing yield gaps

Rapid advancement is made in the field of biotechnology since the discovery of DNA and during subsequent advancements in molecular techniques and other "omics" technologies. This has ushered agriculture into a new era of technological frontiers to tap the latent potential of its biological resources in an unprecedented way, showing a new horizon of opportunities emerge to develop and modernize agriculture. Today, modern agricultural biotechnology encompasses a range of technologies including molecular breeding, fingerprinting, genomics, proteomics, genetic engineering, genome-editing, tissue culture and micropropagation techniques, and other advanced applications. This has empowered scientists, provided unlimited potential, to develop new strategies to harness genetic potentials for solving current and emerging crop production challenges. Therefore, biotechnology has provided a unique capacity to successfully fighting back the continuing battle against diseases, pests, and environmental stresses that are global threats to the survival of mankind. Genetic engineering, a part of modern biotechnology, involves the manipulation of the gene(s) of crop species by introducing, eliminating, or editing specific gene(s) through modern molecular techniques.

During the 1970s and 1980s, the public sector began supporting biotech research with lots of anticipations to advance the use of genetic engineering in agriculture soon to be taken over by the private sector. The first genetically modified (GM) plants were successfully developed as early as 1983 using antibiotic-resistant tobacco and petunia. In 1990, China started to commercialize GM tobacco for virus resistance followed by the Flavr Savr tomato in the United States. By 1995 and 1996, several transgenic crops were approved for large-scale use. Since the first commercial delivery in 1996, millions of smallholder farmers around the world have become beneficiaries of the multiple benefits from growing GM crops [19, 20].

Farmers are primary beneficiaries of the improved production and associated positive environmental, socio-economic, health impacts [21]. The rapid adoption and expansion of biotech crops reflect the substantial multiple benefits realized by farmers in industrial and developing countries. To date, of interest to farmers are several GM crops with enhanced input traits, such as disease (viral, fungal, bacterial) and insect resistance, herbicide tolerance, and resistance to environmental stresses such as drought, improved processing quality, improved product shelf life, and nutrient-enhanced crops available for commercial production.

Recent data [19] shows global acreage of only four biotech crops, corn, soybean, cotton, and canola has reached 190.4 million hectares in 2019 from 1.7 million hectares in 1996, which is on average 7.9 million hectares growth per year impacting crop production and productivity [22]. In recent years, the novel technique of genome-editing (GE) has been developed for targeted genome modification in plants with a high potential of increasing genetic diversity or correcting genetic

Country	GE crops researched, under testing, under approval process and/or approved	Commercialization (year)
Burkina Faso	Cowpea (insect resistance to *Maruca* pest); Bt cotton resistance to insect pest Bollworm) Rice (Resistance to *Xanthomonas oryza)*	Cotton (2008) suspended from production in 2016*
Cameroon	Cotton (stacked insect resistance and herbicide tolerance)	
Egypt	Wheat, Potato, Maize	Commercial production suspended in 2012
Ethiopia	Cotton (insect resistance); Enset (*Xanthomonas* wilt (BXW) resistance), Maize (insect resistance, drought tolerance)	Bt cotton (2018)
Ghana	Rice (nitrogen use efficiency/water use efficiency and salt tolerance), cowpea (insect resistance to *Maruca* pod borer insect pest), Rice	
Kenya	Cotton (insect resistance), Maize (insect resistance, drought tolerance, and stack of insect resistance and drought tolerance), Cassava (brown streak disease-CBSD), Banana (*Xanthomonas* wilt (BXW) resistance), Sweet potato (resistance to sweet potato virus disease), *Gypsophila* flower, *Sorghum* (biofortification)	Bt cotton (2019); Cassava Brown Streak Disease (CBSD) resistant Cassava (2020); Import ban on GM since 2012
Malawi	Banana plantain (bunchy top resistance), Banana (bunchy top disease resistance), Cowpea (insect resistance), Cotton (insect resistance);	Bt cotton (2018)
Mauritius	Sugarcane	
Mozambique	Maize (and stack of insect resistance, drought tolerance), Cotton (insect resistance)	
Nigeria	Cotton (insect resistance), Maize (insect resistance, herbicide tolerance HT Soybeans, Cassava (delayed postharvest starch deterioration), Cassava (Tuber size increase)cowpea (insect resistance to *Maruca* pest), *Sorghum* (biofortification), Rice (nitrogen use, water efficiency, and salt tolerance -NEWEST) Insect resistance and drought tolerance(Maize)	Cotton (2018) PBR Cowpea (2019) Bt Maize (2021)
South Africa	Cotton (insect resistance, herbicide tolerance multi-stack), Maize (insect resistance, drought tolerance, and stack of insect resistance and drought tolerance), Soybean (stacked trait with modified fatty acid composition); sugarcane (insect resistance); Wheat (insect resistance), Potato (insect resistance), Sugar beet, Tomato, Sweet potato, Cucurbits, Ornamental bulbs, Cassava; Apple, Strawberry, Apricot, Peach, Table grapes, Banana (data of traits for these crops has not been obtained).	Bt cotton (1997) Bt- Maize (1998) Bt- & Dt-Maize (2018?) Soybean (2001)
Sudan	Cotton (insect resistance)	Bt cotton (2012)
eSwatini	Cotton (insect resistance)	Bt cotton (2019)
Tanzania	Maize (drought tolerance; stacked for insect resistance and drought tolerance)	
Uganda	Banana (Xanthomonas wilt (BXW) resistance, Black Sigatoka resistance, Pro-vitamin A, Nematode and weevil resistance), Cassava (Cassava mosaic disease virus, Cassava whitefly resistance, Cassava mosaic disease virus, cassava brown streak disease virus resistance), Cotton (Bollworm resistance, herbicide tolerance), Maize (Insect resistance (stemborer), Drought tolerance, Drought tolerance and insect resistance (stacked genes), Rice (Nitrogen use efficiency, salt tolerance, water use efficiency), Sweet potato (Weevil resistance), Soyabean (Herbicide tolerance), Potato (Potato blight resistance).	

Source: ISAAA (2018), ISAAA Biotech Updates (2020), ISAAA Biotech Update (2021).

Table 2.
Genetically engineered (GE) crops researched, under testing, approval or commercialization in different countries of Africa.

defects. The simplicity and high efficiency of these tools have made it optimal for precise genome editing, heralding a new frontier in the—"Gene-revolution"—and in the development of modern biotechnology.

GM technology has been targeting some of the yield constraints and successful technologies have been commercialized in Africa for different crops such as insect resistance (maize, cotton, soybean, brinjal, cowpea), disease resistance (cassava, potato, sweet potato), better nutrition and quality (rice, potato, sorghum, banana). Some of these technologies are now successfully tested or grown in some countries of Africa (**Table 2**). Globally, by the end of 2019, a total of 71 countries (excluding EU countries) [19] issued regulatory approvals for GM crops, of these 11 were African countries. Total approval granted between 1992 and 2018 has reached 4349 from 70 countries (28 countries from EU) for food (2063), feed (1461), and environmental release or commercial cultivation (825) of GM plants [23]. In 2020 alone, 43 approvals were recorded for GM crops globally, involving 33 varieties from 12 countries, and eight of them are new varieties [22]. In 2019, four countries in Africa have given commercially approved for GM crops namely Ethiopia, (Bt cotton), Malawi (Bt cotton), Kenya (Bt cotton), and Nigeria (PBR cowpea) for the first time. Nigeria had additional approval for TELA maize in October 2021 and Kenya approved GM Cassava in June 2021. The TELA maize is built on the progress made from a decade of excellent breeding work under the WEMA project and working toward introducing the Bt- gene to WEMA, water-efficient varieties for drought tolerance [15, 16].

Despite several crops under testing for a long period, only a few have been commercialized in Africa (**Table 2**) [24]. In the SSA, South Africa has taken the lead with an estimated 2.7 million hectares covered with GM crops. It grows three commodities, namely cotton (100% cover), maize (85%), and soybeans (95%) of the total acreage [25]. Nigeria follows with three approvals (Bt cotton, PBR Cowpea, and TELA Maize) since 2018 [17], whereas Sudan stands second in acreage (about 192,000 hectares) from Bt cotton production.

Yield and quality improvements and associated economic benefits of growing GM crops have been the driving factors for biotech crops' rapid global expansion. A study conducted on GM crops and conventional hybrid (CH) maize yield differences across 106 locations and over 28 years in South Africa has shown a mean yield increase for GM over CH maize of more than 0.42 MT per hectare in addition to reducing yield risks [26]. Others reported [27] that GM technology adoption has reduced chemical pesticide use on average by 37%, increased crop yields by 22%, and increased farmer profits by 68%. According to the report, yield gains and pesticide reductions are larger for insect-resistant crops than for herbicide-tolerant crops, and yield and profit gains are higher in developing than in developed countries.

3. Farmers access to new agricultural technologies

Since the first field trial of a GM product back in 1987, the world has seen massive progress in the adoption of biotechnology crops and products and an increasing number of laboratory and field trials for a variety of novel GM products. Of the total global acreage (190 million hectares) of GM crops in 2019, the share of African countries is close to 3.0 million hectares only with South Africa taking the lead with 2.7 million hectares for HR-soybeans, IR/DT- maize and Bt cotton, followed by Sudan for 192,000 hectares of Bt- cotton [21, 28]. Currently, however, 13 biotech crops containing 13 traits in 13 countries are under different stages of research and evaluation in Africa [21]. Crops such as cotton, maize, cowpea, rice, sorghum, potato, sweet potato, cassava, banana, and sugarcane are either at the stage of

Confined Field Trials (CFT) or commercial production status [29]. Since 2018, four countries have entered commercial production for the first time in Africa namely, Nigeria (Bt cotton and PBR cowpea in 2018 and TELA maize in 2021), Kenya (Bt cotton in 2020 and virus resistant cassava in 2021), Ethiopia (Bt cotton in 2018), and Malawi (Bt cotton in 2018), after approval for the respective GM crops [19, 20]. Nigeria has made a move to become the first among African nations followed by Kenya that approved commercial use of GM food crops cowpea and maize.

Given global advancement in the use of GM crops, progress in Africa has been slower than expected [30, 31]. After three decades of global experience on the safety of GM crops and impressive impacts on the livelihood of millions of farmers, many countries still are postponing approvals of GM crops. Numerous health and environmental safety research reports have sufficiently confirmed the safety and desirable impacts of GM crops and their derived products [30–34]. Such scientific evidence have not challenged enough the lingering public perception and controversies around the risks of GM crops [35]. Instead, the overwhelming challenges faced by farmers make it difficult to believe these technologies can positively affect the situation of smallholder farmers [31]. However, scientists believe genetic engineering and genome-editing technologies will continue to impact the global economy with new momentum for more innovative technologies. Countries such as Ghana, Tanzania, Ethiopia, Mozambique, Uganda, and Malawi are in process of working on clarifying the biosafety context and developing a guideline for promoting genome-editing technologies in crop improvement [36].

3.1 Factors shaping access and availability of biotech products for smallholder farmers

The commercialization of already approved products is challenged by a wave of issues along the product commercialization chain. The national research capacity has been very critical to respond to farmers' needs for new technologies through creating awareness to the public, advising policymakers, testing of technologies, approvals, and helping access to proven technologies by farmers. In the same way robust regulatory system is needed to respond to applications based on scientific and empirical evidence. Often this has been a challenge in most countries since sufficient safety data generated can only be accepted and reviewed again by the regulatory agency of adopting country. Private and public sector developers apply step-wise review and decision processes to critically monitor the development of new products and to ensure that only good events are commercialized. Therefore, the intellectual property, product stewardship, and commercialization strategy become key parts of the product life cycle.

The Excellence Through Stewardship (ETS) [37], a global industry coordinated organization, identifies the key steps in the biotechnology product life cycle which includes the following: (i) research and discovery; (ii) product development; (iii) seed or plant production; (iv) marketing and distribution; (v) crop production; (vi) crop utilization; and (vii) product discontinuation (**Figure 4**). Product Stewardship and commercialization are key cross-cutting components along the product life cycle for the industry to remain innovative and viable. Successful commercialization of a GM crops, therefore, requires a well-planned strategy with sufficient information and expertise in a wide range of professions spanning from research and discovery to market and consumer interest.

In other words, success in commercialization also depends on downstream activities: functional seed systems and extension systems, strong technology demonstration, presence of reliable financial and marketing services, and the like. These are often weak in developing countries including most parts of Africa. The

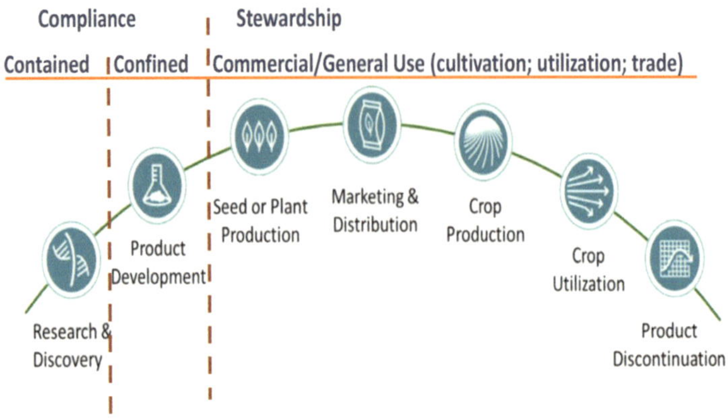

Figure 4.
Biotechnology product life cycle (Excellence Through Stewardship, 2018). Source: Excellence through Stewardship (2018).

blame on lack of political will, safety concern, or public acceptance for the delay in the adoption of deregulated products is often misleading. A recent assessment of stakeholders view on commercialization barriers of released biotech products shows socio-economic constraints, high cost of seed, weak certification of seed, weak private sector involvement, inadequate awareness of the technology, and best practices to be important [18, 24, 38, 39]. Thus, potentially a stronger public-private partnership in research, product development, and product commercialization in developing countries holds the key.

3.2 Challenges of scaling-up and utilization of biotech crops

Rigorous risk assessment studies take years to complete only to satisfy the benefit of the doubt. In Africa, many consider GM crops are intended for use in industrialized countries and are hence inappropriate for agriculture in Africa. There is a poor understanding of the use and potential impact of the technologies on improving productivity. In some countries, GM crops are considered a threat to biodiversity due to fear of replacing local or conventional varieties and indigenous crop species and thereby making farmers dependent on private seed companies. Limited research, regulatory and monitoring capacities, and anticipated loss of export markets with trade-sensitive countries also add up to the challenges against wider commercialization of the biotech crops [38]. In countries that have overcome hurdles of the regulatory system, rolling of GM crop commercialization and access by growers depend much on what happens downstream the pathway beyond product development, regulatory approval, and registration.

3.3 Enhancing regulatory decisions for improved access

Delayed decisions from regulatory agencies have a large, negative impact on the commercialization of new GM crop varieties around the world, but also in Africa [28]. While some delays can be sustained by some private sector developers, public sector developers are reliant on funding cycles and their projects are more quickly discontinued by indecision at regulatory agencies [40]. Regulators can strengthen decision-making by first reviewing the safety of new GM products and then linking the decision to national policy goals such as food security, sustainability, and the economic benefits to local farmers [41]. Linking regulatory decisions on GM plants to national policy goals, such as achieving the UN Sustainable Development Goals

(SDGs), will help to clarify which products benefit the community, the environment, and bring about economic growth [18].

3.4 Seed access

After going through national performance and verifications studies to satisfy national variety release and registration requirements [29], the product deployment is carried out by the technology owner, mostly a private company, through technology demonstration and demand-based seed supply. In this process, roles and stakeholder institutions change where the private sector, seed system, extension system, and other regulatory and financial institutions take over and function in subsequent steps. These transitions are not always clearly defined where the public sector is a major supplier of improved seed or where the seed sector is predominantly informal as in most African countries. Therefore, the commercialization of GM crops is overburdened with multiple issues of promoting new and approved products.

Weak seed systems and weak credit systems limit product access by farmers. A recent study on Bt-cotton hybrid seed access by farmers indicates that weak coordination among various stakeholders along the seed value chain is shown to exacerbate the problem of sustainable supply and wider utilization of the approved GM products [38, 39]. Lack of awareness of role players, inadequate demonstration of new technology to farmers as well as poor handling of the new technology by farmers, and poor extension schemes also contribute to the poor commercialization observed. Socio-economic constraints such as the high cost of hybrid seed, weak certification of seed, and inadequate awareness of technology and best practices (seed handling, agronomy, etc) can become important factors that can slow or block progress in some countries [38]. This also requires a stronger public-private partnership to advance the integration of modern biotechnology in the national R&D system.

4. Country case studies

4.1 Burkina Faso

4.1.1 Country progress

Burkina Faso has signed the Cartagena Protocol on Biosafety in 2003. It has an active and functional regulatory system hosted by the National Biosafety Agency (NBA) (Agence Nationale de Biosécurité, ANB) currently exercising Biosafety laws, regulations, policies, and guidelines in the country. In addition, at a regional level, the Economic Community of West African States (ECOWAS) has put regional framework and rules on biosafety. The NBA is hosted by the Minister of Higher Education but has consultative bodies such us National Scientific committee of Biosafety (comité scientifique national de Biosécurité = CSNB), Scientific and Technique Council, National observatory of Biosafety regrouping members from various ministries and non-governmental organizations.

4.1.2 Product development

The NBA has approved different research activities on GM crops. From 2006 to 2015 about 32 permits for different GM cotton activities related to BollgardII, RRF (herbicide tolerance), and the stack of both were made for import, laboratory studies, CFT, commercialization, and seed production activities. From 2010 to 2021, there were six permits given for *Maruca* Pod Borer resistant GM cowpea using

Cry1Ab or *Cry2Ab* genes for greenhouse and CFT. Other GM crop permits provided include for CFT on Bt Maize for insect resistance; greenhouse trial for vitamin and zinc-rich biofortified sorghum; greenhouse trials for leaf blight resistance in rice.

Only the Bt cotton Burkina Faso had reached the stage of commercialization and utilization. However, the Bt cotton cultivation was discontinued in 2016 due to cotton fiber length issues associated with the marketing of Bt cotton. Currently, most of the research activities are carried out in the greenhouses, cages, and CFTs. In Burkina Faso, stakeholders support the use of GMO as a solution to food security and for human disease control such as Malaria. The ANB has been undertaking sensitization of various public entities and various stakeholders since 2009 on biosafety actions as described by the national legislation and the Cartagena Protocol.

4.2 Ethiopia

4.2.1 Country progress

Ethiopia signed the Convention on Biological Diversity (CBD) in 1993, Cartagena protocol in 2000 which was approved by Parliament in 2003. The country adopted a tighter regulatory framework based on the Precautionary Principle (equivalent to "No GMO") ratified in 2009. The Biosafety bill was debated amended in 2016, known as 'A Proclamation to Amend the Biosafety Proclamation 2009'. In 2017, the National Biosafety Advisory Committee was adopted and in 2018 the country issues its Biosafety Guidelines. The amended law permitted scientists and institutions to do research and education pertaining GMOs. This allowed to establish legal and regulatory systems and build technical capacity to support and manage GMO issues and approved after CFT of three Bt cotton varieties in 2016 under the procedure of "Special permit", a provision in the Biosafety Law for research purposes. This was followed by 2 years of NPT across seven sites until 2018. The country approved two Bt cotton hybrids, JKCH-1050 and JKCH-1947 originally obtained from JK Agri Genetics Ltd., India for environmental release and variety registration. The accelerated commercial release demonstrated Ethiopia's government commitment to support the cotton development to satisfy booming textile industries [29].

4.2.2 Product development

Ethiopia considered biotechnology as one of the priority areas in its National Science and Technology Policy formulated in 1993 [42]. Due to interest to tighten the non-GMO stand, the prohibitive regulatory system delayed its overall engagement in modern biotechnology, postponed the use of available products, and hampered the development of the local capacity building. After approval of two Bt cotton Bollgard I type varieties in 2018, demand for Bt cotton seed for 2021/22 estimated at 3250 kg was requested for 1300 hectares. Some level of cross-border Bt cotton seed also takes place with Sudan and around 3055 hectares around border areas are already covered with such imported Bt cotton seed.

In 2008, the Biosafety Authority and the NBAC granted a "Special Permit" approval for CFT of drought-tolerant (WEMA) and insect resistant (TELA) maize for testing from 2018 to 2023. The isogenic conventional lines were evaluated for 2years in different locations before the CFT. The two-year CFT was started in 2019 under a controlled drip irrigation system for drought-tolerant trait evaluation and has shown very promising results. The stacked maize environmental release for both insect resistance and drought tolerance is awaiting approval using existing provisions.

In 2013, Ethiopia deployed GM technology for its indigenous Enset crop (also called "false banana") improvement in collaboration with the International Institute of Tropical Agriculture (IITA) for developing varieties with resistance to the deadly bacterial wilt disease caused by *Xanthomonas campestris* [43]. The collaborative research work had begun at BecA, Nairobi at IITA laboratory and later moved to Holetta Agricultural Biotechnology Research Center (NABRC) in Ethiopia in 2018 after approval was obtained for contained use (Contained Lab Permit). Approval for testing transgenic Enset under CFT is underway. Further to its endeavor in GM technologies, Ethiopia will soon engage in testing Late Blight Resistant (LBR) resistant cisgenic potatoes. Application submitted for CFT is awaiting approval.

4.2.3 Farmers access to new agricultural technologies

Approved Bt cotton hybrid seed demand is increasing but the hybrid Bt cotton seeds are not locally available and need to be imported from the technology supplier. But due to the decline in exports during the COVID-19 Pandemic, the Bt cotton seed supply system has suffered from foreign exchange restrictions to purchase seeds. The absence of local seed companies investing in Bt cotton seed has been one of the key challenges facing Bt cotton commercialization in Ethiopia.

Stakeholders across the cottonseed system must assess the most feasible pathway to ensure easy access to quality seeds at a reasonable cost, especially to smallholder farmers. Supporting cotton production with appropriate extension services and training of farmers and other relevant stakeholders for best practices is required to scaling-up the use of Bt cotton in the country. Developing innovative partnerships with technology developers to enable local Bt cotton hybrid seeds production will help to achieve affordable and sustainable access to GM technology.

4.2.4 Public perception and acceptance of GMOs

There is no clear data concerning the changes in the public acceptance of GM technologies in Ethiopia. However, the transition at policy and political levels is remarkable; from a stance of "GMO free" advocacy to one with pragmatic consideration to taking advantage of changes and prospects at the global level. The public perception is expected to evolve considerably due to growing global biotechnology importance in promoting food security in the wake of climate change. However, the recent movement following a report by the USDA that recognizes Ethiopia's commitment to implementing the amended protocol and embarking on some GM crops, has sparked severe criticisms against GMOs development in the country [44]. There has been a steep rise in anti-GMO comments following the USDA announcement [45]. It requires to provide the right information to the public and creating the right and positive public perceptions to help the right policy measures and institutional function with respect to biotechnology products.

4.3 Kenya

4.3.1 Country progress

Kenya is among the first African countries that signed the Cartagena Protocol on Biosafety in 2002. It also set up a national biosafety regulatory authority followed by a Biosafety policy signed into law in 2010 [46]. The exercise of dealing with GM products has seen many challenges such as the one when the government through the Ministry of Health instituted a Moratorium on the import and trade of GMOs on November 21, 2012, an embargo that remains in force to this day [47].

To date, two crops have been approved for commercialization use in Kenya and these are the Bt cotton hybrid, which was commercialized in 2020, and the improved cassava variety for resistance to Cassava Brown Streak Disease (CBSD). The NBA approved the application for environmental release for GM cassava containing Event 4046 in 2021 [48]. The GM cassava has increased root quality and higher yields [49]. Kenya is the first country globally to consider a request for environmental release involving GM cassava crops. Many other crops are now at different stages of regulatory approval. In the year 2021, 36 applications have been submitted for various crops under review [48].

4.3.2 Farmers access to new agricultural technologies

Kenya's GMO regulatory framework is robust and active. It is designed for regulating contained use, import, export and transit, environmental release, and labeling [46]. The emerging research area of gene-editing technologies in food and agriculture presents the newest frontier in the area of legislation and regulations in Kenya [46]. The NBA board has undergone timely training to equip them with knowledge on the understanding of the regulatory process of genome-edited organisms and products in Kenya [46].

4.3.3 Challenges in product commercialization

A strict and arduous regulatory approval framework remains one of the most important challenges to GMO adoption in Kenya [50]. So far, Bt cotton has been commercialized and the status of Bt-maize is at the NPT stage. Access to Bt cotton hybrid seeds, access to credit to purchase Bt cotton seeds, and lack of adequate monitoring data for Bt cotton is the weak side of the commercialization process.

Among the public institutions, Government Counties can play a role by forming cotton-producing clusters to support access to Bt cotton hybrid seed and inputs and access to the cotton market to encourage cotton-producing smallholders. This exercise on Bt cotton can also be helpful for similar efforts in the future for other new technologies [51].

4.3.4 Public perception and acceptance of GMOs

Public perception of GMOs in Kenya has been mostly negative for a long time due to bad press and negative publicity about GM products [50]. Kenya had instituted a moratorium on GMO import and trade in 2012 based on a study by Séralini et al. [52] that has since been disapproved. The damage, however, had been done and slowed progress in GM acceptance and adoption in the country. For most of the public, GMOs were dangerous, and disposed the government to take a reactive action. The growing awareness on the benefits of GMO technology in the continent and in Kenya in particular, is seeing an upsurge in attitude change for the better [50].

4.4 Malawi

4.4.1 Country progress

Malawi has made significant progress in biotechnology and biosafety since the ratification of the Cartagena Protocol on Biosafety in 2009. The country has domesticated the protocol by developing a legal and institutional framework for biosafety. Malawi developed its Biosafety Act in 2002, Biosafety Regulations in 2007, and enacted Biotechnology and Biosafety Policy in 2008. The CFT and NPT Guidelines, Trial Manager Handbook, and Inspectors Handbook were prepared in 2007. Since 2009, three permits to conduct GM crop trials have been issued under the Biosafety Act and approved its first Bt cotton for commercialization in 2018.

Other GM crops initiatives were transgenic Banana and Bt Cowpea both of which were terminated in 2019 due to lack of finance to support the research.

4.4.2 Farmers access to new agricultural technologies

Malawi's biosafety legal framework does not hinder the commercialization of approved technologies. Before varietal release of the Bt cotton hybrids, field demonstrations across key cotton-growing districts were done to help farmers with the potential of the technology (Bollgard II) and hybrid cotton varieties to help farmers build a positive perception about the benefits. However, the cost of Bollgard II hybrid cotton seeds was US$30 (MK 25,000) in 2021 became a concern. This means that for a hectare, farmers spend US$ 123.5 at a seeds rate of 4 kg/ha compared to US$ 1.2/kg for OPVs. The Bt cotton seed grown in Malawi are supplied from India and transport/import cost make seed prices higher and affects the adoption of the technology by smallholder farmers. Trainings on GM cotton seed multiplication for local farmers is underway to reduce cost on seed importation which is anticipated to result into affordable seed cost and improve its accessibility and adoption by smallholder farmers.

4.4.3 Public perception and acceptance of GMOs

In Malawi issues such as biosafety concerns, public acceptance, political will, and support influence the adoption of GM crops. Public opinion has not been contradicting to the introduction of GM cotton possibly due to the absence of known negative impacts on human health and good publicity during the field demonstration trials. There is high political will as government is working to restore the cotton industry in the country. Regulatory decisions have been science-based and risk assessment is done on case-by-case basis which has built level of trust for the technology among farmers and the public.

4.5 Nigeria

4.5.1 Country progress

Modern biotechnology regulation in Nigeria started in the early 1990s. The Convention on Biological Diversity (CBD), which Nigeria signed in 1992, identified GMOs or LMOs as a group of organisms produced by modern biotechnology that needed special attention because of their perceived adverse impacts on biodiversity and human health. Based on the Convention's recommendation, Nigeria ratified its biosafety framework in 2002. Consequently, research practice began in modern biotechnology, along with it the biosafety legal regime became apparent. Subsequently, Biosafety Law was put in place in April 2015 giving birth to the National Biosafety Management Agency (NBMA) for the implementation of the Act which also became amended in 2019.

4.5.2 Progress in product development

To keep abreast with advancements in modern biotechnology, Nigeria developed several guidelines including for GM Food, Feed Processing, GM Mosquito, GM Trees, Birds, Fish, and other animals. The country is the first in Africa to validate Genome editing guidelines during the last quarter of 2020. Several processing permits were granted for food and feed from GM maize, soybeans, and others.

Currently, Nigeria has several R&D activities at different levels: research, testing, pipeline, and commercialization. To date, NBMA has approved CFTs for the following crops: Bio-fortified cassava enhanced with pro-vitamin A, iron, and zinc; GM cassava resistant to cassava mosaic virus, Cassava brown streak disease virus, and enhanced with iron and zinc. Also, cassava was modified for higher starch; cowpea modified for resistance against maruca, HT soybeans; GM rice modified for nitrogen use efficiency, water use efficiency, and salt tolerance and GM maize for resistance to stem borer/fall armyworm and drought tolerance. The approval for commercial release has been for GM cotton (Bollgard II) to Bayer Agriculture Nig. Ltd./Mahyco Agriculture Private Ltd. in July 2018; cowpea modified for resistance to maruca insect pest and insect-resistant/drought-tolerant maize (TELA).

4.5.3 Farmers access to new agricultural technologies

The most important regulatory constraints are related to finance and laboratory facilities. The challenge in product commercialization of GM crops, as experienced in cowpea, is meeting the seed demands of farmers. Whereas in the case of cotton, the cost of seeds is not affordable by smallholder farmers, concerted efforts are being made by various platforms such as the open forum on agricultural biotechnology (OFAB), in Africa, Nigeria Chapter in collaboration with extension agents to let farmers get the right information and advisory services on biotechnology products. Nigeria's Biosafety Law requires mandatory labeling of products containing GM products or ingredients exceeding 4%, which restricts market access for GM products.

4.5.4 Possible pathways for commercialization

Access to improved seed is realized when the farmers can buy the seeds when they need them at an affordable price. Trust building is critical so that farmers as pragmatic as they are, have a positive attitude toward GM technology despite anti-GM campaigns and their misconceptions.

4.5.5 Perception and acceptance of GMOs

The Nigerian public has a mixed opinion about GM crops and their food products due to mixed information about the importance of biotech in promoting food security and the public concerns about its safety and health-related issues. A higher number of the public in Nigeria believe the country should domesticate the technology and build local capacity to develop GM crops [53]. For example, policymakers' and scientists' perception on GM technology was examined in Ghana and Nigeria using semi-structured interviews [54]. Results showed most respondents including policymakers believe the technology has great potential to solve agricultural problems. However, lack of trained personnel and weak institutional capacities present significant challenges to its wider utilization.

4.6 Sudan

4.6.1 Country progress

Sudan is a member of the Cartagena Protocol on Biosafety (CPB) since 2005. In 2010, a national biosafety law dealing with the application of modern biotechnology was issued and in 2012, Biosafety Council was formed. However, biosafety measures are only partially in place for the implementation of the Cartagena Protocol [55]. Despite such efforts by the government to develop the biosafety regulatory system, much remain to be done for the effective implementation of the protocol on biosafety [56]. The national biosafety law was amended to become

"Miscellaneous Amendments Law" (Unification of Environment Councils) and officially gazetted in Sudan [57].

The first open-pollinated Bt cotton genotype (CN-C02) carrying Bt gene Cry 1A from which is a specific toxin against larvae of bollworm was introduced by China-aid Agricultural Technology Demonstration Center (CATDC) and released for commercial production under the name Seeni1 in 2012. The Seeni1 variety was fast adopted at a commercial scale from 19,300 hectares in 2012 to 61,300 hectares in 2013 [58]. In 2016, the area almost doubled to 120,630 hectares. Seeni1 occupied about 25% of the country's total cotton cultivation area in 2012 and 97% in 2014 [59]. After the successful adoption of the first Bt cotton variety, Seeni1, another open-pollinated Bt cotton genotype from China (SCRC37) carrying the same gene of Seeni1 was released for commercial production and named Seeni2 in 2015. In the same year, two Indian Bt cotton hybrids; JKCH1947 (Hindi1) and JKCH1050 (Hindi2) carrying JKAL X-gene (Cry1Ac), were also released for commercial production [60]. The area under Hindi2 progressively increased from 7560 hectares to 33,600 hectares in 2021. The total Bt cotton cultivated area in Sudan since first commercial production in 2012 has grown to occupy about 98% of the total cotton area in 2021. In Sudan, cottonseeds represent a valuable oil and cake source. The major concern after the Bt cotton commercialization is the food safety of its byproducts; however, permissible levels for GMOs intended for direct use as food/feed needs approval from the national biosafety committee.

Recently transgenic cotton hybrid varieties carrying Cry1AC + Cry2A and glyphosate-tolerant trait (CP4 ESPS) were approved by the national biosafety technical committee in compliance with the national biosafety regulations for further testing. In Sudan, the establishment of national action plans for developing and promoting cotton exports and harmonizing its marketing policies are seen as crucial steps to restore Sudan's position in the international cotton market.

4.6.2 Farmers access to new agricultural technologies

In Sudan, Bt cotton is the only GM crop under commercial production since 2012. Additional new transgenic cotton varieties approved by the national biosafety committee are under testing and will enrich the Bt cotton variety options. The national seed industry of transgenic crops is not fully complying with the biosafety regulations due to the limited awareness of stakeholders involved in the seed industry. This has caused the sub-standard seed to be distributed by dealers.

Almost all Bt cotton seeds for open-pollinated variety are produced by the private seed sector under the governance of public institutions. The current situation of seed production could be improved with policy to guide and incentivize seed producers (public and private) for high-quality seed supply. The trend of seed demand growth in Sudan has been clear since Bt cotton adoption and requires comprehensive situation analysis to install a visionary seed production scheme.

On the other hand, not all smallholder farmers can access good quality seed because of limited financial support and a lack of farmers' organizations to obtain agricultural credit. Enabling policies are required for smallholder cotton farmers to overcome this problem and related marketing challenges.

4.6.3 Public perception and acceptance of GMOs

Sudanese public participation in GMOs use debates and its general awareness is limited. Either lack of understanding or misperception of the technology predominates. Public-wide formal and informal education on safety concerns (biosafety and food safety) and GMO utilization need to be strengthened. More engagement and participation of stakeholders along the cotton value chain would help to have a clear plan for promoting and sustainability utilizing the products of GM technology. Currently, the adoption of transgenic cotton in Sudan is farmer-driven and government intervention is highly beneficial to strengthen farmers' associations for market access and improving the benefits of Bt cotton to local farmers.

4.7 Uganda

4.7.1 Country progress

For the past 15 years, Uganda has been steadily integrating biotechnology into national development processes and developing local capacity. The Uganda national biotechnology strategy identified biotechnology as a tool to address challenges in the agricultural sector [61, 62]. The government has been providing support to build human resources and research infrastructure capacity to strengthen research development and innovation in biotechnology and played a dominant role in Uganda. R&D using modern biotechnology tools in crop science was initiated in 2003 at the National Agricultural Biotechnology Center. Other institutions like Makerere University and the National Agricultural Research Organization's (NARO) followed suit to join the effort. Several international and regional organizations also have been supporting national crop biotechnology R&D including USAID, Bill and Melinda Gates Foundation, ASARECA, CIMMYT, and Rockefeller Foundation. Through support from the government and development agencies, more than 10 research laboratories have been established for biotechnology research and development. The scientific community in Uganda has embraced biotechnology and is actively engaged in R&D using modern biotechnology and genetic engineering tools. There has been a growing application of tissue culture, molecular diagnostic tools, and the development of genetically engineered transgenic crops.

4.7.2 Biosafety regulatory system

Uganda ratified the Cartagena Protocol on Biosafety in 2001 [63]. In 2008, the government of Uganda adopted the National Biotechnology and Biosafety Policy to provide a regulatory and institutional framework for the safe and sustainable application of biotechnology for national development. Uganda's biosafety institutional framework includes national competent authority, national focal point, the national biosafety committee, monitoring and compliance mechanisms, and institutional biosafety committees.

The Uganda National Council for Science and Technology (UNCST) serves as the national competent authority and provides regulatory oversight for GMO research and development programs through the National Biosafety Committee (NBC). To support the NBC, biotechnology research institutions have established Institutional Biosafety Committees (IBC) to provide research biosafety stewardship and serve as a link between the research scientists and NBC. To provide a comprehensive biosafety regulatory framework for commercialization of GM crops, the Parliament of Uganda introduced the Genetic Engineering Regulatory Bill in November 2018 to be assented into an act. The Bill was seconded through stakeholder policy consultations to ensure establishment of an enabling national biosafety legislation.

4.7.3 Country progress

The first field trial of GM crops was conducted in 2007 on genetically engineered bananas for resistant to Black Sigatoka disease. To date, the NBC has approved 17 field research trials involving several GM crops mentioned below for various crops and traits (**Table 2**) [64–66]. The detailed summary of GM crops and incorporated traits is also partly presented in **Table 2**.

Like other breeding product pipelines, GM products require on-farm agronomic and agroecological tests under the guidance of approved biosafety guidelines. In Uganda, scientists are unable to proceed with product testing on farmer's fields to ascertain GM product performance due to a lack of national biosafety legislation and regulations. Crops such as banana (research, CFT and multilocation trials), Cassava (CFT, multi-locational trials), Cotton (CFT, multi-location trials), Maize (CFT and multi-location trials), Rice (CFT Research), Sweet potato (Greenhouse), Soybean (Greenhouse), Potato (CFT- Multilocation trials) have not been tested on farmers fields. Research on these crops has been conducted through joint collaborations involving local and international institutions such as NARO, IITA, AATF, Queensland University of Technology (QUT), Leeds University, Donald Danforth Plant Science Center (DDPSC), Bayer, International Potato Center (CIP), Makerere University, and Michigan State University.

5. Lessons learned and future prospects

Biotechnologies can help African country's efforts toward achieving social and economic development and contributing to the United National (UN) Sustainable Development Goals (SDGs) through improving agricultural productivity and increasing resilience to climate change impacts. As highlighted in the six case studies, countries in Africa are at various stages of biotechnology R&D and regulatory capacities. With the recent positive decisions made by the governments of several countries in Africa, the future holds prospects for the commercialization of GM products. Research, regulatory, and outreach capacity in modern biotechnology is seen as fundamental to the promotion of advanced science and technology in research programs including GMO and genome-editing research and development.

Identifying policy and regulatory gaps and adjusting to meet current and future needs would always be required to promote agricultural biotechnology for sustainable development in biotech and non-biotech countries. Proactively working toward building awareness of stakeholders and right public perception and relentless effort to capacitate policymakers would help to maintain the current efforts in improving political dynamics toward modern biotechnology and avoid sliding back to the old rhetoric led by postmodernist anti-GMO and anti-technology activism.

Since it took several years of negative publicity to entrench distrust among the public, it can only be undone with unyielding and consistent communication and outreach espousing, especially positive benefits to smallholder farmers and consumers and farmers as champions. Therefore, strong voices are necessary to champion the adoption of GMOs and genome-editing technologies in countries in Africa. Misinformation and disinformation, and competing interests inevitably complicate how modern biotechnology is viewed and its benefits are harnessed in Africa for smallholder farmers. The science communication should be amplified with messaging centering around a farmer and consumer benefits and contributions to UN Sustainable Development Goals (SDGs).

The transitions from product development to deployment and commercialization are often difficult in developing countries. Multiple institutions from the public and private sector including the farming communities are involved to operate. This needs to be well aligned and coordinated institutional functions are needed to ensure sustainable access and deployment of new technologies/products by smallholder farmers while keeping product integrity, quality, and excellence through stewardship. Experience shows the importance of careful handling and management of new technology with simultaneous preparation for the local seed systems to ensure that new products are consistently available and affordable by smallholder farmers. Alternative technologies are needed for widening the scope of adoption through a healthy market and avoiding negative perceptions to impinge on efficiency and competitiveness.

Farmers are willing to adopt impactful technologies that can enhance agricultural productivity and their livelihoods. However, closer consultation and understanding of their challenges is critical to foster and sustain repeated adoption of GM crops by farmers to convey a realistic understanding of the production and marketing challenges and receive necessary policy support. A clear monitoring strategy is needed for field management of GM crops and their sustainable use and impacts as well as co-existence in the farming systems of adopting countries.

Author details

Endale Gebre Kedisso[1]*, Nicolas Barro[2], Lilian Chimphepo[3], Tahani Elagib[4], Rose Gidado[5], Ruth Mbabazi[1], Bernard Oloo[6] and Karim Maredia[1]

1 Michigan State University, USA

2 National Biosafety Agency, Burkina Faso

3 Environmental Affairs Department, Malawi

4 Agricultural Research Corporation (ARC), Sudan

5 National Biotechnology Development Agency (NABDA), Nigeria

6 Egerton University, Kenya

*Address all correspondence to: kedissoe@msu.edu

References

[1] AGRA Annual Report. 2017

[2] Jayaram K, Riese J, Sanghvi S. Agriculture: Abundant Opportunities. McKinsey Quarterly; 2010 https://www.mckinsey.com/featured-insights/middle-east-and-africa/africas-path-to-growth-sector-by-sector

[3] FAO and World Bank. Derived from Farming Systems and Poverty. Food and Agriculture Organization of the United Nations (FAO) & World Bank; 2006 Global Map of Irrigation Areas, FAO, NRLW & Johann Wolfgang Goethe University, Frankfurt am Main, Germany GAUL [Accessed: August 28, 2021]

[4] Sasu DD. Prevalence of food insecurity in sub-Saharan Africa as of 2019, by level. Statista 2021; https://www.statista.com/statistics/1190797/food-insecurity-in-sub-saharan-africa/ [Accessed: August 8, 2021]

[5] Baquedano F, Cheryl C, Kayode A, Jayson B. International Food Security Assessment, 2020-30, GFA-31. U.S. Department of Agriculture, Economic Research Service; 2020

[6] Nyiawung RA, Suh N, Bishwajit G. Trends in Cereal Production and Yield Dynamics in Sub-Saharan Africa Between 1990-2015. Journal of Economic Impact. 2020;**1**(3):98-107

[7] Anthony VM, Ferroni M. Agricultural biotechnology and smallholder farmers in developing countries. Current Opinion in Biotechnology. 2011. DOI: 10.1016/j.copbio.2011.11.020

[8] FAO. In: Elcio P, Guimaraes EP, Ruane J, Scherf BD, Sonnino A, Dargie JD, editors. Marker-Assisted Selection. Current Status and Future Perspectives in Crops, Livestock, Forestry and Fish. Rome: FAO; 2007

[9] FAO. The Special Challenge for Sub-Saharan Africa. High Level Expert Forum—How to Feed the World in 2050. Viale delle Terme di Caracalla, Rome, Italy; 2009

[10] Agricultural Science and Technology Indicators – ASTI, Boosting Investment in Agriculture Research in Africa Building. A Case for Increased Investment in Agricultural Research in Africa. Report Commissioned By Au-Safgrad; 2021

[11] UNECA. Towards Achieving the African Union's Recommendation of Expenditure of 1% of GDP on Research and Development. Addis Ababa: © UNECA; 2018 https://hdl.handle.net/10855/24306

[12] Eicher CK, Maredia K, Sithole-Niang I. Crop biotechnology and the African farmer. Science Direct Food Policy. 2006;**31**:504, www.elsevier.com/locate/foodpol-527

[13] National Academies of Sciences, Engineering, and Medicine. Preparing for Future Products of Biotechnology. National Academies Press, 2017; MLA 8th Edition (Modern Language Assoc.).

[14] Gert-Jan S, Beintema N. An Assessment of The Critical Financial, Human, and Institutional Capacity Issues Affecting Agricultural Research in West Africa: Synthesis and Policy Considerations. Background document for The World Bank. Washington, DC: IFPRI; 2017

[15] African Agricultural Technology Foundation (AATF). Strong supportive voices for biotechnology crucial for technology uptake in Africa. 2021. Available from: https://www.aatf-africa.org/strong-supportive-voices-for-biotechnology-crucial-for-technology-uptake-in-africa/

[16] Oloo B, Maredia K, Mbabazi R. Advancing adoption of genetically

modified crops as food and feed in Africa: The case of Kenya. African Journal of Biotechnology. 2020;**19**(10):694-701. DOI: 10.5897/AJB2020.17159

[17] ISAAA AfriCenter. 2021

[18] Gebre EK, Maredia K, Guenthner J, Koch M. Commercialization and deployment of genetically modified crops in Africa: Opportunities and challenges. African Journal of Biotechnology. 2022. DOI: 10.5897/AJB

[19] ISAAA. Accomplishment Report 2020

[20] ISAAA. Brief 55: Global Status of Commercialized Biotech/GM Crops. 2019

[21] ISAAA. Brief 54: Global Status of Commercialized Biotech/GM Crops. 2018

[22] Zhang J. Overview of global GMO development. AgroNews Annual Review. 2020

[23] Ichim MC. The more favorable attitude of the citizens toward GMOs supports a new regulatory framework in the European Union. GM Crops & Food. 2021;**12**(1):18-24. DOI: 10.1080/21645698.2020.1795525

[24] Turnbull C, Lillemo M, Hvoslef-Eide TAK. Global regulation of genetically modified crops amid the gene edited crop boom—A review. Frontiers in Plant Science. 2021;**12**:630396. DOI: 10.3389/fpls.2021.630396

[25] Venter C. Quick overview of GMOs in South Africa. AgriOrbit. 2020

[26] National Academy of Science. Chapter: 6 Social and economic effects of genetically engineered crops. 2016. In: Genetically Engineered Crops: Experiences and Prospects. National Academy Press; 2021

[27] Klümper W, Qaim M. A meta-analysis of the impacts of genetically modified. Crops. 2014. DOI: 10.1371/journal.pone.0111629

[28] Aldemita RR, Hautea RA. Biotech crop planting resumes high adoption in 2016. GM Crops Food. 2018;**9**(1):1-12. DOI: 10.1080/21645698.2018.1428166

[29] Akinbo O, Obukosia S, Ouedraogo J, Sinebo W, Savadogo M, Timpo S, et al. Commercial release of genetically modified crops in Africa: Interface between biosafety regulatory systems and varietal release systems. Frontiers in Plant Science. 2021. DOI: 10.3389/fpls.2021.605937

[30] Gbashi S, Adebo O, Adebiyi JA, Targuma S, Tebele S, Areo OM, et al. Food safety, food security and genetically modified organisms in Africa: A current perspective. Biotechnology & Genetic Engineering Reviews. 2021. DOI: 10.1080/02648725.2021.1940735

[31] Azadi H, Samiee A, Mahmoudi H, Jouzi Z, Khachak PR, De Maeyer P, et al. Genetically modified crops and small-scale farmers: main opportunities and challenges. Critical Reviews in Biotechnology, Early Online. 2015:1-13. DOI: 10.3109/07388551.2014.990413

[32] EUROPEAN COMMISSION (EU). A Decade of EU-funded GMO Research (2001-2010). 2010; https://www.researchgate.net/publication/233770770

[33] ISAAA Crop Biotech Update; 2021

[34] Bayer. GMO & Biotechnology Unlocking Hidden Potential. 2020; https://www.cropscience.bayer.com/innovations/seeds-traits/gmo-biotechnology.

[35] Mbabazi R, Dissanayake HG, Guenthner J, Maredia K. Perception of international stakeholders on genetically modified organisms (GMOs). Asian Biotechnology and Development Review. 2016;**18**(2):51-57

[36] Thompson T, Gyatso T. Technology adoption for improving agricultural productivity in Sub-Saharan Africa. In: 021 Virginia Tech College of Agriculture and Life Science and Virginia Tech Foundation. 2020 Available from: https://globalagriculturalproductivity.org/technology-adoption-for-improving-agricultural-productivity-in-sub-saharan-africa/

[37] Excellence Through Stewardship (ETS). Guide for Stewardship of Biotechnology Derived Plant products. Washington DC: Excellence Through Stewardship; 2017

[38] Mbabazi R, Koch M, Maredia K, Guenthner J. Crop biotechnology and product stewardship. GM Crops & Food. 2021;**12**(1):106-114. DOI: 10.1080/21645698.2020.1822133

[39] Alhassan SW, Anthony VM, Rudelsheim P. Implementation of stewardship in African agricultural public research and development institutions. Journal of Agricultural Biotechnology and Sustainable Development. 2018;**10**(5):97-103

[40] Morris and Thomson. Genetically modified crops commercialized in South Africa. In: Wambugu, Kamanga, editors. Biotechnology in Africa: Emergence, Initiatives and Future. Springer; 2014

[41] Raybould A. Improving the Politics of Biotechnological Innovations in Food Security and Other Sustainable Development Goals. Transgenic Res: Springer; 2021

[42] Waktola A, Tsegaye B. Biotechnology related policy, management and negotiation competence: Case Study from Ethiopia. In: NORAGRIC, Center for International Environment and Development Studies. Norwegian University of Life Sciences, Ås. Noragric Report No. 14-B. 2003

[43] IITA Blogs. Project to Save Enset—Ethiopia's Ancient False Banana from Deadly Bacterial Wilt Disease Star. IITA.; 2013

[44] Sida T. Will Ethiopia be a springboard or a stonewall for GM crops in Africa? Nature Biotechnology. 2021;**2021**(39):147-148. DOI: 10.1038/s41587-021-00827-5

[45] USDA Foreign Agricultural Services. Agricultural Biotechnology Annual Report. Ethiopia. Report Number: ET2019-0010; 2020

[46] Biosafety News. 2021. Available from: https://www.biosafetykenya.go.ke/images/newsletter2021/NBA-BIOSAFETY-NEWS-5TH-EDITION.pdf

[47] USDA-FAS. Agricultural Biotechnology Annual; Agricultural Biotechnology, Country: Kenya; Report Number: KE2020-0008Annual. 2020. Available from: https://apps.fas.usda.gov/newgainapi/api/Report/DownloadReportByFileName?fileName=Agricultural%20Biotechnology%20Annual_Nairobi_Kenya_10-20-2020 [Accessed: September 25, 2021]

[48] National Biosafety Agency. NBA Board Approves Environmental Release Application for GM Cassava. Press Release. 2021. Available from: https://www.biosafetykenya.go.ke/images/NBA-Board-Approves-application-for-Genetically-Modified-Cassava-FINAL.pdf [Accessed: September 25, 2021]

[49] XINHUANET News. Scientists Bank on GM Crop to Lift Cassava Industry in East Africa. 2021. Available from: http://www.xinhuanet.com/english/africa/2021-07/06/c_1310044498.htm [Accessed: September 25, 2021]

[50] Maina J. Bad press and wild claims have unfairly slowed GM crops on the continent, African Scientists say. Alliance for Science (cornell.edu). 2021

[51] Silas DO, Maredia K, Kedisso EG, Guenther J and Jeremy O. Value-chain of

cotton industry in Kenya with focus on product stewardship for timely provision of certified quality seeds to farmers. (In preparation).

[52] Séralini GE, Clair E, Mesnage R, Gress S, Defarge N, Malatesta M, et al. Long term toxicity of a Roundup herbicide and a Roundup-tolerant genetically modified maize. Food and Chemical Toxicology. 2012;**50**:4221-4231

[53] USDA. Agricultural biotechnology. Nigeria approves the commercial release of Bt. Pod-Borer Resistant Cowpea. 2019:1-13

[54] Adenle AA. Stakeholders' perceptions of GM technology in West Africa: Assessing the responses of policymakers and scientists in Ghana and Nigeria. The Journal of Agriculture and Environmental Ethics. 2014. DOI: 10.1007/s10806-013-9462-y

[55] Sudan Fourth National Report under Cartagena Protocol on Biosafety. 2019

[56] Ali AM. Biosafety and biosecurity in countries with low resources—Sudan as an example biosafety. 2013, 2013;**2**:1. DOI: 10.4172/2167-0331.1000105

[57] Gazette No. 1091: Miscellaneous Amendments Law—(Unification of Environment Councils) of 2020—Legislation No.3 of 2020

[58] Naglaa AA. The story behind Bt cotton: Where does Sudan stand. GM Crops & Food. 2014;**5**(4):241-243. DOI: 10.1080/21645698.2014.997119

[59] Ali AM. et al. Submission of Application to the National Biosafety Council, Ministry of Environment, for the Approval of the release into the Environment of a Bt Cotton Variety Seeni, 2012

[60] Ali AM., National Variety Release Committee Meeting, A Proposal for the Release of an Open Pollinated Bt Cotton (*G. hirsutum*) Variety for Production in the Sudan, 2015

[61] Ecuru J, Naluyima H. Biotechnology developments in Uganda and associated challenges. African Crop Science Journal. 2010;**18**(3):133-139

[62] Agaba J. GM Crops: Kenya and Nigeria Progress as Uganda Falters. 2021; https://allianceforscience.cornell.edu/blog/2021/08/gm-crops-kenya-and-nigeria-progress-as-uganda-falters/

[63] Muheebwa H. Uganda Parliament Passes Bill To Promote Use Of Genetically Modified Materials, Biotech. For Intellectual Property Watch 2. International IP Policy News. 2017. Available from: https://www.ip-watch.org/2017/11/03/uganda-parliament-passes-bill-promote-use-genetically-modified-materials-biotech/

[64] Zawedde BM, Kwehangana M, Oloka HK. Readiness for environmental release of genetically engineered (GE) plants in Uganda. Frontiers in Bioengineering and Biotechnology. 2018;**6**:152. DOI: 10.3389/fbioe.2018.00152

[65] Afedraru L. Uganda's scientists strive to use biotechnology to solve agricultural, health and environmental challenges. Genetic Literacy Project. 2018 https://geneticliteracyproject.org/2018/05/31/ugandas-scientists-strive-to-use-biotechnology-to-solve-agricultural-health-and-environmental-challenges/

[66] Muhumuza JR. Uganda's Biotech Research: More Than Just Untapped Potential 2021; https://allianceforscience.cornell.edu/blog/2021/03/ugandas-biotech-research-more-than-just-untapped-potential/

Section 2

Biosafety Framework and Consumer Perceptions

Chapter 7

Proposed Revision of the National Gene Technology Scheme for Australia

Robert Redden

Abstract

Plant breeding was provided access to wider genetic variation through genetic modification (GM) of crops in the 1980s. This involved transfer of DNA between species, and introduction of new traits into domestic crops. Concerns were raised for the outcomes in food health and in the environment with GM crops, with the spectre of 'Frankenstien' foods and fear of the unknown. This led to widespread adoption of GM regulations based on the 'Precautionary principle' of safeguarding the risks to health and to the environment, even when scientific evidence was lacking to support these concerns. The Green lobby required GM foods to be safe for consumption, with no ill-effects over the long term and for many generations into the future. GM foods have proven safe for over two decades, and with benefits to crop productivity, pest and disease resistances, improved nutrition and tolerances of extreme climatic stresses. GM includes the new biotechnology of Genome Editing (GE), with targeted and precise changes to gene sites, and inter-specific transfer of genes from poorly accessible Crop Wild Relatives (CRW), for adaptation of crops to climate change. Food and fibre crops need to be exempt from GM regulations.

Keywords: Regulation, genetic modification, genome editing, crops, climate change, crop wild relatives

1. Introduction

As outlined by Redden [1], Australia's cultivation of GM crops in 2015 comprised herbicide-tolerant canola 444,000 ha, stacked GM (herbicide-tolerant plus pest resistant) cotton 253,000 ha, and herbicide tolerant only cotton 20,000 ha [2, 3].

With GM cotton pesticides have been substantially reduced, benefiting human safety, adjacent livestock enterprises and the environment, plus improving yields [4–6]. Herbicide resistant canola both controlled weeds and raised yields [5, 7]. These GM crops can be grown with minimum tillage, thereby conserving soil moisture for crop maturation in the low rainfall Southern cropping zone where every mm saved is 20 kg/ha or more grain [8]! Herbicide weed control allows earlier sowing to better match crop growth with seasonal winter rainfall.

South Australia (SA) was the last mainland state to have a moratoria on GM crops [1]), scheduled to 2025 but now lifted as recommended by Anderson [9]. The moratoria cost the canola industry $33 million over 2004–2018. Australian GM canola with a 10% yield benefit, suffered no adverse international market advantage compared with non-GM canola except for Japan, which paid an estimated price premium of $32/tonne (about 7%) for GM free (zero adventitious contamination) canola from Kangaroo Island (KI) in SA [9]. This entailed segregation of non-GM from GM canola in the delivery-chain with identity protocols and codes of practice. The moratoria was kept for KI crops, and the market chain for KI produce will remain segregated.

In Tasmania GM crops have been banned since 2001 [10]. This is supported by the horticulture and honey industries maintain Tasmania's image for pure GM free produce.

2. Issues

2.1 Regulation of GM crops in Australia

The National Gene Technology Scheme (NGTS) in Australia was enabled by the Gene Technology Act 2000. Regulation is administered by the Office of the Gene Technology Regulator (OGTR), to apply a process based 'Precautionary' approach to any kind of directed genetic alteration [1, 11], specifically DNA transfer between species.

The object of the Act for all living organisms is: *'To protect the health and safety of people, and to protect the environment, by identifying risks posed by gene technology, and by managing those risks through regulating 'dealings' with GMOs'.*

OGTR authorises the release of GM crops in coordination with other agencies; Food Safety Australia and New Zealand (FSANZ), the Australian Pesticides and Veterinary Medicines Authority, Therapeutic Goods Administration, National Industrial Chemical Notification and Assessment Scheme, Department of Agriculture and Water Resources, and Department of the Environment and Energy [11, 12].

CRISPR Genome Editing (GE) is able to alter genetic expression without transfer of new genetic material with the SDN1 procedure, as a more advanced version of GM. OGTR has made a recent incremental change to a 'Principles based' flexible approach, with recognition of the SDN 1 with a product history of low risk [4, 11, 12]. However OGTR risk assessment and oversight remain, plus the regulations of complementary agencies.

SDN 1 genome editing is classified as GM/GE under 'Notifiable Low Risk Dealings' (NLRD) [11, 12]. NLRD products cannot be released to the environment without OGTR approval, and must be compliant with OGTR regulations for transport, storage and disposal, while GM field trials have to be registered and isolated [11]. NLRDs must be approved by the Institutional Biosafety Committee (IBC) and OGTR [11, 12]. Costs apply for administration, risk assessment and management.

OGTR requires that GM/GE crop development must undergo detailed case-by-case assessment of risks to food safety and to the environment, with research and development conducted in contained facilities; this is expensive research [13, 14]. This is based on the 'Precautionary' principle, rather than 'Outcome' based with recognition of benefits to society and the environment.

The science of gene technology is poorly understood publicly, enabling the Green lobby to demonise GM for socio-economic reasons or to challenge details of a scientific study [15, 16], or now to raise fears that SDN 1 GE is GM in disguise, so allowing GM foods to be unlabelled and hidden from the public [17, 18].

The anti-GM lobby is well funded in USA through tax deductions to 'organic' and environmental groups [19, 20]. Anti-GM protesters have destroyed GM field trials in UK and Australia, and with non-scientific health and environmental claims supported risk regulation of GM crops and discouraged developing countries from approving GM crops [18, 19, 21, 22]. Organic certification demands no GM products, so that the organic industry has a large vested interest in denigrating GM.

Foods derived from GM crops pose no greater safety risk than from conventional plant breeding [2, 4, 5, 23, 24]. GM food safety has been validated with over 25 years of research by the American Medical Association [25], World Health Organisation [26], The British Royal Society [27], and 500+ independent institutions. GM crops benefit the environment primarily by substantially reducing the use of toxic pesticides/fungicides [28].

2.2 Genome editing (GE)

The new GE techniques such as CRISPR enable precise changes to the genome, with cutting of DNA at a specific location, and insertion, deletion, or modification of nucleotides in a gene, and include gene silencing, gene enhancement, and synthetic genes (**Figure 1**) [4, 29, 30].

China has heavily invested in GE with the purchase of Syngenta [31]. Genome editing has been developed for tomato, potato, maize, rice, wheat, sorghum and citrus, and presents a major challenge to GM crop regulators [4]. GE dramatically increases the number of traits which can be modified in crops, in a manner which is far quicker and cheaper than the original GM technology has been able to achieve [30, 32].

Base pair alteration (SDN 1) may be indistinguishable from either a random mutation or what may be achieved by conventional breeding, and is regarded as very low risk for health and the environment [4]. It is unlikely however to replace most uses of GM from before 2010 and already in farmers' fields. The SDN 2 CRISPR procedure involves larger DNA changes with a DNA repair template, while

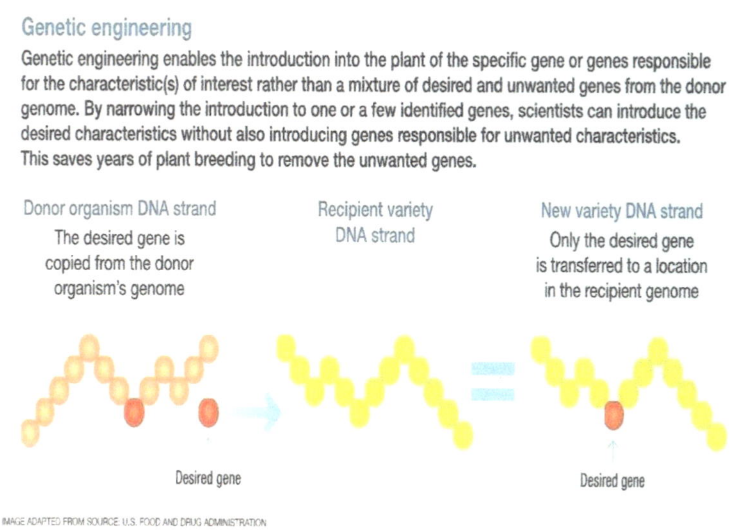

Figure 1.
Image adapted from source U.S. Food and Drug Administration [29].

SDN 3 enables targeted insertion of foreign DNA, both are still subject to full OGTR regulation.

The CRISPR-Cas9 DNA insertion is displayed in **Figure 2** [29].

Occurrences of 'off-target' changes are very rare in plants and detectable by whole genome sequencing [4]. Mutation breeding has always been exempt from regulations, a precedent for SDN 1 GE.

2.3 GM regulation and GE

Policies on GM regulation are evolving with changes in biotechnology, but at different rates and to different extents in various countries. Genome editing targets the introduced traits themselves rather than the technology used to create them, in contrast to the traditional process-triggered GM regulatory system championed by Europeans [13, 33, 34]. EU does not exempt GE from GM regulations [4, 35].

In recent national responses to advances in GE [36]; USA, Norway, Australia, New Zealand, Japan, and Argentina either permit SDN 1 genetic changes, or are considering relaxation of regulation. Lassoued *et al.* [37] reviewed plant breeders on deregulation of GE, who noted increased ease of transformation, gain in precision, and improved opportunity to introduce novel traits from Crop Wild Relatives (CWR) through GE. Public education about GE was seen as necessary, plus opportunity for public participation in legislative processes to relax regulations [38]. New regulatory frameworks have been proposed [39–42], with the latter suggesting a product based approach for regulation of GE crops, especially now that genome sequencing is complete for over 200 plants and under development for over 10,000 genome assemblies.

Agribio Victoria can process 50,000 SNPs at a time, and has sequencing capabilities for reliable detection of interactions between large numbers of different genes. These affect the majority of traits of agricultural interest, and can be a significant complement to the expression of major genes such as 'blackleg' resistance in canola [43]. The advances in sequencing and in GE together make possible the targeted transfer of complex abiotic stress tolerance traits from CWR to domestic crops.

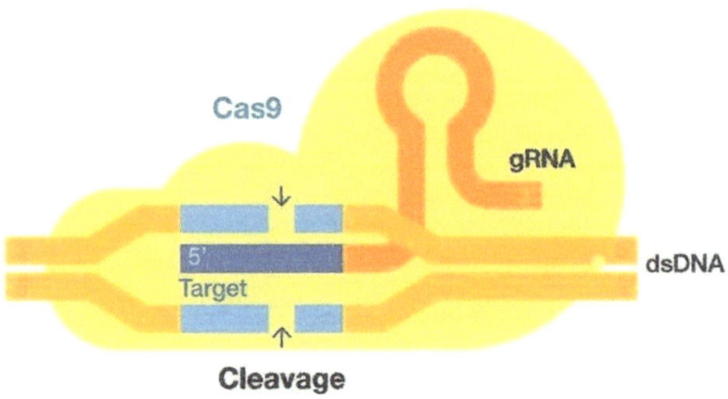

Figure 2.
A Schematic diagram of the Cas9 enzyme (yellow) and the guide RNA (gRNA) that directs the enzyme to cleave double-stranded DNA (dsDNA) at specific sites. Image adapted from source: Marus Walter, Attribution-share alike 4.0 International (CC BY-SA 4.0).

However worldwide acceptance of revised regulations would be needed to achieve international consensus and removal of asynchronous trade barriers [44–46], which are significant barriers to international commercialisation of GM/GE [47, 48].

Future challenges include a warmer more variable climate for which CRW can provide genes for abiotic and biotic stress tolerances [40, 41, 49]. In many cases biotechnology applications can assist introgression of these stress tolerant traits into crops [4]. This would help to address twin challenges to agriculture of climate change and food security for a predicted 10 billion people by 2060 [50].

2.4 Climate change and genetic adaptation

World food security has become severely threatened since the introduction of regulations on gene technology for crops over 20 years ago [1]. Gene technology regulation needs to recognise that crop environments are becoming more variable and challenging. There has been an unprecedented growth in world population by over three-fold in the last 100 years to 7.85 billion today [50], with an equally dramatic 60% rise in the greenhouse gases, especially CO2 mainly from coal, oil, gas and cement sources of pollution to over 400 ppm [51], resulting in a continual but fluctuating increase in global mean temperature towards 1.5°C above pre-industrial levels since 1900 [51]. On most scenarios this warming will rise above 2°C by 2100, with the lowest emission scenario very unlikely to eventuate, with increasing urbanisation and more energy intensive life styles. Certain trends such as polar warming can set up reinforcing feedback loops for warming: ice melts, permafrost thaws, and desertification. Spikes in high temperature will be from a higher base, and frosts and droughts will be more severe especially upon seed set. Food security will be under threat [30, 49].

Thus a climate crisis for agriculture has intensified since the 1990s, when genetic modification of food and fibre crops raised safety concerns. However GM crops have been shown to be beneficial with improvements in crop and food nutrition, disease and pest resistances, yield productivity, and tolerances of drought, high temperature, frost and salinity [4].

Now in the 2020s there is an urgent need to widen the genetic diversity of food and fibre crops to address the coming challenges of abiotic and biotic crop stresses with Climate Change [30, 41, 49]. GE provides the tools to exploit the largely untapped genetic diversity of CWR, the evolutionary ancestors of crops [17, 30], with precise introgression of genes for abiotic/biotic tolerances (heat, frost, and drought tolerances, salinity, pest and disease resistances). CWR have genetic diversity for adaptation to far more extreme environments than crops were exposed to during domestication over the past 12,000 years, and provide opportunities to transform crop adaptation to Climate Change [17, 52]. However it is an immense challenge to implement GE transformations across all crops; from vegetables, spices, cereals and legumes to root crops and fruits, before the world is stranded with agricultural systems un-adapted to changed environments.

There is a future opportunity cost in not recognising that climate change combined with an unprecedented growth in population creates an urgency to re-adjust GM regulation, to promotion and acceptance of new gene technologies, especially GE [16, 29, 45, 53–55]. NGTS can re-align towards an aspiration of crop adaptation (climate proofing) to climate change [24, 39]. Advances in cropping ingenuity and crop genetics will be essential to produce more food in more hostile environments.

2.5 Proposal for a revised NGTS for food and fibre crops only, in Australia

An appropriate tiering of regulation for crops should recognise outcomes of product benefits to farming and the environment, and a long established food safety record.

GM/GE food and fibre crops should be exempt from NGTS regulation [1]. The current NGTS/OGTR over-regulation stifles the opportunity to realise the benefits from CWR for adaptation to climate change, raises costs, and tends to exclude GM/GE research and development from small research organisations. The present costs to market for GM/GE crops are prohibitive [16]. The current NGTS/OGTR regulations are no longer fit for purpose, and NGTS could be changed to exempt food and fibre crops only, but not vaccine and pharmaceutical crops, microorganisms and animals [1, 11].

A Revised NGTS [1] for food and fibre crops would have a new aim: '*Genetic improvement of food and fibre crops by application of gene technologies, with recognition of product outcomes of agricultural, health and environmental benefits*'.

This Revised NGTS would greatly reduce operational costs of the plant-centric OGTR and better secure its funding sustainability, without the monitoring, surveillance and compliance activities for GM/GE food and fibre crops.

A restructured OGTR could change from regulating GM food and fibre crops, to play a major role in educating the public on the benefits of new biotechnologies with publications, educational webinars and social media posts [1]. OGTR has the required expertise to explain and illustrate new developments in biotechnology [11, 12]. This could be supported with championing of a Revised NGTS for food security in a more populous world with a changing climate.

3. Summary of a proposed revision of NGTS for crops in Australia

OGTR regulations on GM food and fibre crops need to be removed for equivalence with conventionally bred crops. The proposal is for an exemption of GM food and fibre crops from current NGTS regulation, and adoption of a Revised NGTS for sustainability of agriculture under climate change.

As proposed by Redden [1], a Revised NGTS would include:

- Regulations of relevant agencies such as OGTR, FSANZ, and APVMA, to be science based and supportive of GM products.

- Exemption of GM food and fibre crops from NGTS/OGTR legislation, yet still comply with FSANZ standards.

- Deployment of current and new gene technologies for world food security, even as cropping environments become less favourable.

- Research organisations to champion the introgression of genes from CWR into crops for improved productivity, food nutrition, and adaptation to abiotic and biotic stresses.

- An education campaign across primary to tertiary education levels, and social media.

- Risk objections to GM crops and derived foods to be science based, taking into account both medical expertise on health risks and social and environmental benefits.

- Relaxation of regulations for crop GE would facilitate new market entrants for GM crops and broaden the scope of GE across more crops and key traits.

- Individual food choice is retained, but labelling requirements should not be burdensome on GM derived foods.

- International trade barriers to GE produce are removed as other countries also rollback GM regulations on food and fibre crops.

- Co-existence of GM and non-GM crops is manageable in Australia, given existing SA segregation protocols and stack management practices at grain reception points.

- GE also benefits the organics industry, both with genetic resistances to pests and diseases, and tolerances of abiotic stresses.

- Excluded from the proposed Revised NGTS are vaccine, and pharmaceutical crops, micro-organisms and domestic animals.

Author details

Robert Redden
RJR Agricultural Consultants, Horsham, Victoria, Australia

*Address all correspondence to: bbredden@yahoo.com.au

References

[1] Redden, R. 2021. An opinion paper: Genetic modification for Agriculture; proposed revision of GMO regulation in Australia. Plants 10 (in Press)

[2] Brookes, G. 2016. Adoption and impact of genetically modified (GM) crops in Australia: 20 years' experience. PG Economics Report prepared for CropLife Australia Ltd, Canberra,

[3] WA. 2019. Information on genetically modified organisms. . Government of Western Australia, Food and Agriculture Newsletter, 13 August, 2019.

[4] ABCA (Agricultural Biotechnology Council of Australia). 2020. The official Australian reference guide to Agricultural Biotechnology and GM crops. Fourth edition. Agricultural Biotechnology Council of Australia. https://www.abca.com.au/

[5] Brookes, G. 2020. Crop biotechnology continues to provide higher farmer income and significant environmental benefits. PG Economics Limited. www.pgeconomics.co.uk

[6] Naranjo, S.D., Ellsworth, P.C. 2009. Fifty years of the integrated control concept: moving the model and implementation forward in Arizona. Pest Management Science. https://doi.org/10.1002/ps.1861

[7] ACIL Tasman. 2007. GM Canola: An Information Package. Department of Agriculture Fisheries and Forestry, with the assistance of Innovation Dynamics. ISBN 978-0-957221-8-3.

[8] Sadras, V.O., Angus, J.F. 2007. Benchmarking water-use efficiency of rainfed wheat in dry environments. Field Crops Research 100 (2-3), 125-138.

[9] Andserson, K. 2019. Independent Review of the South Australian GM Food Crop Moratorium. Report to the SA Minister for Primary Industries and Regional Development.

[10] Breen, F. 2019. Tasmania's GMO ban good news for some, a 'missed opportunity' for others. Tas Country Hour, 8 August 2019 at 6:41am.

[11] NGTS (The Third Review of the National Gene Technology Scheme. 2018. © 2018 Commonwealth of Australia as represented by the Department of Health.

[12] OGTR (Office of Gene Technology Regulation). 2020. Technical Review of the Gene Technology Regulations 2001. Discussion paper: Options for regulating new technologies. http://www.ogtr.gov.au/internet/ogtr/publishing.nsf/Content/reviewregulations-1

[13] Friedrichs, S., Takasu, Y., Kearns, P., Dagellier, B., Oshima, R., Schofield. C., Moreddu, C. 2019. An overview of regulatory approaches to genome editing in agriculture. Biotechnology Research and Innovation 3: 208-220.

[14] Nicolia A, Manzo A, Veronesi F, Rosellini D. 2013. An overview of the last 10 years of genetically engineered crop safety research. Crit. Re. Biotechnol. 2013 Sep 16.

[15] Gennaro, A. 2020. European Food Safety Authority rebuffs activist criticism of its safety assessment of insect and herbicide resistant GMO crops. European Food Safety Authority July 8, 2020.

[16] Jaffe, G. 2016. Genetically Engineered Foods and their Regulation: the Way Forward after Twenty Years of Adoption. Regulatory Focus. August 2016. Regulatory Affairs Professionals Society.

[17] Haijar, R., Hodgkin, T. 2007. The use of wild relatives in crop improvement: A

survey of developments over the last 20 years. Euphytica: 1-13.

[18] Druker, S. 2014. The UK's Royal Society: a case study in how health risks of GMO's have been systematically misrepresented. Independent Science News for Food and Agriculture.

[19] Genetic Literacy Project 2020. https://geneticliteracyproject.org/2020/07/08/european-food-safety-authority-rebuffs-activist-criticism-of-its-safety-assessment-of-insect-and-herbicide-resistant-gmo-crops/

[20] Gusovsky, D. 2014. Big lobbying dollars flow in engineered food fight. AGRICULTURE, CNBC. Published Tue, Jul 29 2014 4:29 AM.

[21] Lynas, M. 2013. Lecture to Oxfords Farming Conference. Environmental News and Comment, January 3, 2013.

[22] Smyth, S.J. 2017. Genetically modified crops, regulatory delays, and international trade. Journal Food and Energy Security (Jan). https://www.onlinelibrary.wiley.com/doi/full/10.1002/fes3.100

[23] Freedman, D.H. 2013. Are Engineered Foods Evil? Scientific American 309(3): 80-85. doi: 10.1038?scientificamerican0913-80

[24] Macnaghten, P., Haberts, M.G.J.I. 2020. Breaking the impasse: Towards a forward-looking governance framework for gene editing with plants. Plants, people, Planet. 00: 1-13. DOI: 10.1002/ppp3.101107

[25] AMA (American Medical Association), 2012. American Medical Association: GMO labellimg not necessary. https://www.huffpost.com/entry/gmo-labeling-ama-american-medical-association_n_1616716

[26] WHO (World Health Organization). 2016. Frequently Asked Questions on Genetically Modified Foods. https://www.who.int/foodsafety/areas_work/food-technology/faq-genetically-modified-food/en/ Accessed 26 August 2020.

[27] The Royal Society. 2018. Genetically modified (GM) plants: Questions and answers. https://royalsociety.org/topics-policy/projects/gm-plants/

[28] Brookes, G., Barfoot, P. 2016. GM crops: global socio-economic and environmental impacts 1996-2016. PG Economics Ltd, UK.

[29] AAS (Australian Academy of Science). 2019. Genetic modification / Questions and answers. ISBN 978 0 85847 671 4. https://www.science.org.au/

[30] Zhang, Y., Massel, K., Godwin, I.D. & Gao, C. 2018. 'Applications and potential of genome editing in crop improvement', Genome Biology 19, 210, doi:10.1186/s13059-018-1586-y)10.1186/s13059-018-1586-y).

[31] Cohen, J. 2019. To feed 1.4 billion, China gets big on genome editing of crops. https://www.sciencemag.org/author/jon-cohen

[32] Godwin, I. 2020. Gene editing opens a faster and cheaper way to introduce new crop traits. Queensland Alliance for Agriculture and Food Innovation, University of Queensland. https://qaafi.uq.edu.au/article/2020/04/gene-editing-opens-faster-cheaper-way-introduce-new-crop-traits

[33] EU (European Commission). 2010. European Research Area. A Decade of EU-Funded GMO Research (2001-2010). Project Information. https://ec.europa.eu/research/biosociety/pdf/a_decade_of_eu-funded_gmo_research.pdf. Accessed 21 July 2016.

[34] Stokstad, E. 2020. United States relaxes rules for biotech crops. Science and policy, Science, AAAS.

[35] Coghlan, A. 2014. Relaxing of rules urged for genetically modified crops. : $https://www.newscientist.com/article/dn25222-relaxing-of-rules-urged-for-genetically-modified-crops/#ixzz6Uy5EaITQ

[36] Mackelprang, R. 2018. Organic farming with gene editing: An oxymoron or a tool for sustainable agriculture?. mythja/Shutterstock.com University of California, Berkeley.

[37] Lassoued, R., Macall, D.M., Hesseln, H., Phillips, P.W.B. Smyth, S.J. 2019. Benefits of genome-edited crops: expert opinion. Transgenic Res. 28:247-256. (https://doi.org/10.1007/s11248-019-00118-5)

[38] Miles, S., Ueland, Ø., Frewer, L.J. 2005. Public attitudes towards genetically-modified food. British Food Journal 107(4). ISSN: 0007- 700X. https://doi.org/10.1108/00070700510589521

[39] Scheben, A., Edwards, D. 2018. Bottlenecks for genome-edited crops on the road from lab to farm. Genome Biology 19:178

[40] Thompson J, 2002. Genes for Africa: genetically modified crops in the developing world. UCT Press.

[41] Thompson J, 2006. GM crops. The improved potential. CSIRO publishing.

[42] Voytas, D.F., Gao, C. 2014. Precision Genome Engineering and Agriculture: Opportunities and Regulatory Challenges. PLoS Biol 12(6): e1001877. doi:10.1371/journal.pbio.1001877

[43] Cogan, N. 2021. Australian Crop Breeding Research Benefits from Advanced Sequencing with Consistent Methodologies Supporting Australian crop breeders with high-throughput genetic analysis. https//:ilmnmkt.illumina.com/2016006215

[44] Frisvold, G., Reeves, J. 2015. Genetically modified crops: International trade and trade policy effects. International Journal of Food and Agricultural Economics 3 (2): 1-13. Special Issue. ISSN 2147-8988, E-ISSN: 2149-3766.

[45] Menz, J., Modrzejewski, M., Frank Hartung, F., Wilhelm, R., Sprink, T. 2020. Genome Edited Crops Touch the Market: A View on the Global Development and Regulatory Environment. Front. Plant Sci., 09 October 2020 | https://doi.org/10.3389/fpls.2020.586027

[46] Schmidt, S.M., Melinda Belisle, M., Frommer, W.B. 2020. The evolving landscape around genome editing in agriculture Many countries have exempted or move to exempt forms of genome editing from GMO regulation of crop plants. EMBO reports. DOI 10.15252/embr.202050680 | EMBO Reports (2020) 21: e50680

[47] ISAAA. 2018. Global Status of Commercialized Biotech/GM Crops: 2018. ISAAA *Brief* No. 54. ISAAA: Ithaca, NY.

[48] Parisi, C., Tillie, P., Rodríguez-Cerezo, E. 2016. The global pipeline of GM crops out to 2020. Nature Biotechnology 34(1): 31-36.

[49] Coyne, C.J., Kumar, S., Wettberg E.J.B., Marques, E., Jens D. Berger, J.D., Redden, R.J., Ellis N.T.H., Jan Brus, J., Zablatzká, L., Petr Smýkal, P. 2020. Potential and limits of exploitation of crop wild relatives for pea, lentil, and chickpea improvement. Legume Science. 2020;e36. https://doi.org/10.1002/leg3.36

[50] Hickey, L.T., Hafeez, A.N., Robinson, H., Jackson, S.A.,

Soraya C.M.L-B., Tester, M., Gao, C., Godwin, I.D., Ben J. Hayes, B.J., Wulff, B.B.H. 2019. Breeding crops to feed 10 billion. Nature Biotechnology 37: 744-754.

[51] IPCC, 2021: Summary for Policymakers. In: Climate Change 2021: The Physical Science Basis. Contribution of Working Group I to the Sixth Assessment Report of the Intergovernmental Panel on Climate Change [Masson-Delmotte, V., P. Zhai, A. Pirani, S. L. Connors, C. Péan, S. Berger, N. Caud, Y. Chen, L. Goldfarb, M. I. Gomis, M. Huang, K. Leitzell, E. Lonnoy, J.B.R. Matthews, T. K. Maycock, T. Waterfield, O. Yelekçi, R. Yu and B. Zhou (eds.)]. Cambridge University Press. In Press.

[52] Van Tassell, D.L., Tesdell, O. et al., 2020. New crop domestication issues in the age of genome editing: genetic, Agronomic and Cultural change remain co-evolutionary entangled. Front. Plant Sc, 11: 789.

[53] Cavaliero, L.2020. Year 10 Biology, Genetic Modification. Library and innovation centre. https://library.plc.wa.edu.au/year10/biology/GM/genetic_modification

[54] Dewey, C. 2017. The government is going to counter 'misinformation' about GMO foods. Washington Post, May 4, 2017. https://www.washingtonpost.com/news/wonk/wp/2017/05/03/the-government-is-going-to-try-to-convince-you-to-like-gmo-foods/

[55] FAO/WHO. 2000. Safety aspects of genetically modified foods of plant origin. Report of a Joint FAO/WHO Expert Consultation on Foods Derived from Biotechnology. World Health Organization, Headquarters, Geneva, Switzerland. 29 May – 2 June 2000.

Chapter 8

Biosafety Aspects of Genetically Modified Crops

Ashutosh Kumar, Banshidhar, Priyanka Jaiswal and Harmeet Singh Janeja

Abstract

With the advancement in the field of agricultural biotechnology, many genetically modified crops like Bt- cotton, Bt- brinjal have been developed and commercialised to fulfil the need of the world population. Several biosafety concerns viz., risk to human health, risk to environment, ecological concern o has been raised after the rapid commercialization of GM crops every year across the world. As per Convention on biodiversity (CBD), Biosafety is a term used to describe efforts to reduce and eliminate the potential risk resulting from biotechnology and its product. Though many concerns being raised time to time, strict biosafety guideline must be followed before introducing a GM crop in public domain especially in resource poor developing countries.

Keywords: biosafety, GMO, Bt-cotton, CBD, health risk

1. Introduction

GM crops are one of the noble invention of 21st century that holds a good promise for better survival of humanity. These crops are developed through genetic engineering by altering the genetic make-up of the crops for enriching it with one or several economically important traits such as improved quality traits, reduction in anti-nutritional factors, herbicide tolerance, resistance to various biotic and abiotic stresses, etc. The GM crops have helped mankind to stand against various challenges arising out of high population growth, biodiversity loss and climate change but the process following which these crops have been developed may posed serious threat to the biodiversity which serve as the repository of raw materials for various biotechnological applications ranging from improved and processed foods, fibres and fuels, noble medicines and drugs, enzymes, etc. thus it is imperative that the biodiversity must be preserved satisfactorily to fully exploit the potential of this indispensable technology. In recent times, Biotechnological tool such as genetic engineering and recombinant DNA technology has proved its worth in achieving the sustainable development goals and enjoyed a great potential to mitigate the impact of climate change as well and opened new avenues for climate smart agriculture. However, while doing so we must take care of the ultimate stakeholder whether for the biodiversity or the technology i.e. the human beings and its environment. Fulfilling all these contradictory demands concurrently requires an elaborative and exhaustive framework involving robust protocols regarding safe designing, production, handling and transfer of GM crops. Keeping this in view, a series of meeting were held internationally to discuss the possible innovation or strategies to reduce the ill-effects

IntechOpen

of these technological interventions and to develop effective strategies for conservation and preservation of biological resources. One of the practical outcome of these discussion fruits in form of "The Cartagena Protocol on Biosafety, 2000" [1, 2].

2. The Cartagena protocol on biosafety

The Cartagena Protocol on Biosafety (CPB) was adopted on 29 January 2000 in Montreal with the holistic approach to addresses the probable threats from the transfer, handling and use of living modified organisms (LMOs) under the umbrella of Convention on Biological Diversity, 1992 (**Figure 1**).

The term "Biosafety" describes the principles, procedures and policies to be adopted to ensure the environmental and personal safety. The convention directs its Contracting Party to take appropriate measures to regulate, manage or control the risks that may arise due to use and handling of LMOs that may pose some threats to biological and to ensure the safe handling, transport and use of LMOs. Recognising the need of biosafety in genetic engineering research, the Cartagena Protocol on Biosafety (CPB) was adopted with the following objectives:

1. To set up the procedures for safe trans-boundary movement of LMOs.

2. To harmonise principles and methodology for risk assessment and establish a mechanism for information sharing through the Biosafety Clearing House (BCH) [3, 4].

2.1 Guidelines of CPB

The CPB promotes biosafety through well-defined guidelines for the safe transfer, handling and use of LMOs or GMOs, with a specific focus on regulating transboundary movements of these organisms. These guidelines ensure comprehensive information to take decisions on scientifically sound risk assessments and on the precautionary approach in use of LMOs and/or GMOs.

2.2 India's initiative on biosafety

In India, Ministry of Environment & Forests (MoEF) plays the role of the nodal ministry for implementation of Cartagena Protocol and undertakes several initiatives to meet its obligations to the Protocol. It also organised various capacity building programmes to strengthen of the regulatory framework, particularly on transboundary movement of LMOs or genetically modified organisms, risk assessment and its management, training and human resource development and information sharing.

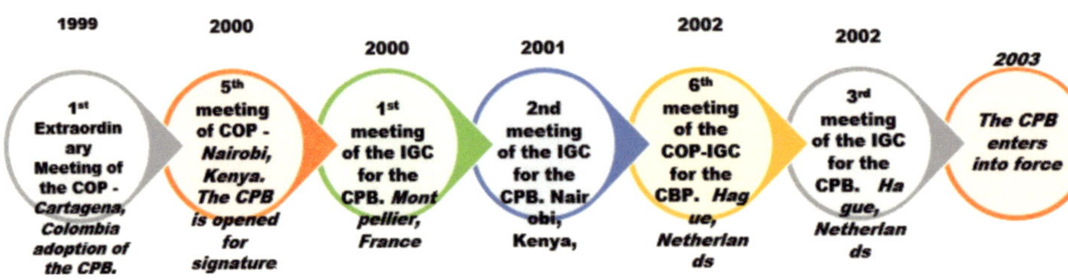

Figure 1.
Timeline of "The Cartagena Protocol on Biosafety". Key: COP: Conference of the parties; CPB: Cartagena protocol on biosafety; IGC: Intergovernmental committee.

Authorities for implementation of regulations and guidelines in the country [5]

1. Recombinant DNA Advisory Committee (RDAC)

2. Review Committee of Genetic Manipulation (RCGM)

3. Genetic Engineering Approval Committee (GEAC)

4. Institutional Biosafety Committees (IBSC)

5. State Biosafety Coordination Committees (SBCC)

6. District Level Committees (DLC)

3. Risk assessment of GM crops

Risk assessment identifies potential hazards and/or adverse impacts of GM crops or derived product on non-target organisms and/or environment. This involves a number of coordinated steps like risk identification, risk characterisation and risk categorisation. The first and foremost practice i.e. risk identification involves identification of risk or possible hazard to the non-target species or the environment, if any, associated with release and use of transgenic or GM crops and associated products. This is followed by overall characterisation of risk i.e., whether its effect are direct or indirect, chronic or acute, immediate or delayed in action, etc. Finally, risk categorisation is done which involvves grouping of identified and well characterised risk under various categories *viz.*, negative health effects on target population; adverse effect on non-target population, the evolution of resistance or resurgence in the targeted pest/pathogen population, flow of transgene to another species, etc. In the process of risk assessment if a potential risk is identified the appropriate measures are taken for its management .

3.1 Steps of risk assessment

4. Risk management in use of GM crops

Risk management involves strategic techniques to reduce the adverse effect of GM crops and associated products on non-target species or environment and also to reduce the chances of development of resistance in target pest population. Several tactics *viz*. application of alternate or mix insecticides with different modes of action or use of refuge strategy could be effectively employed to minimise the risk of development of insecticidal resistance in insects. These techniques are also helpful in avoiding the problem of resurgence in insects. In Bt crops newer techniques *viz*. use of alternate or combined Bt toxin or refuge strategy are much rewarding in management of resistance. Similarly, weeds could develop resistance in them following various mechanism *viz*. modified site of action, detoxification and compartmentalisation. By doing so they rendered the herbicides or weedicides ineffective against them in long run. Thus, to minimise or to prevent the risk of development of herbicide resistant in weeds and evolution of super weed various techniques have been utilised. Rotation of herbicides or using them in combination effectively reduces chances of development of resistance against herbicides in weeds. Crop rotation is another technique that could be used to reduce this risk [6].

5. Conclusion

Modern advances in biotechnology has revolutionalised the way of living particularly in meeting the requirement of food, fodder, fibre and fuel by use of GM crops. However, a group of social activist and environmentalists are always in against of the use of GM crops because of its unprecedented effects on ecosystem and human health. Thus, a scientific debate has been continued for a long time in which the favouring statements are made based on risk assessment and its consecutive management as per the norms and protocols. In addition to this a number of initiatives have already been taken by national as well as international agencies to ensure safety measures in use, handling and transfer of these GM crops. Thus, basically these crops can be commercialised in public domain with adequate care following the defined biosafety measures.

Author details

Ashutosh Kumar*, Banshidhar, Priyanka Jaiswal and Harmeet Singh Janeja
Department of Genetics and Plant Breeding, College of Agriculture, Lovely Professional University, Phagwara, Punjab, India

*Address all correspondence to: ashutoshgpb@gmail.com

References

[1] Bawa AS, Anilakumar KR. Genetically modified foods: Safety, risks and public concerns—A review. Journal of Food Science and Technology. 2013;**50**(6):1035-1046

[2] Hilbeck A, Meier M, Römbke J, Jänsch S, Teichmann H, Tappeser B. Environmental risk assessment of genetically modified plants-concepts and controversies. Environmental Sciences Europe. 2011;**23**(1):1-12

[3] https://www.cbd.int/doc/publications/bs-brochure-02-en.pdf

[4] https://www.icgeb.org/wp-content/uploads/2019/01/2-The-Cartagena-Protocol-on-Biosafety-Julian-Kinderlerer.pdf

[5] https://ibkp.dbtindia.gov.in/DBT_Content_Test/CMS/Guidelines/20181115134719867_Regulations-Guidelines-for-Reocminant-DNA-Research-and-Biocontainment-2017.pdf

[6] Prakash D, Verma S, Bhatia R, Tiwary BN. Risks and precautions of genetically modified organisms. In: International Scholarly Research Notices. 2011

Chapter 9

GM Crops: The West versus the Rest

Jennifer Thomson

Abstract

This chapter will explore the reasons why some countries, broadly described as "developed," do not allow their farmers to plant GM crops. It will then go on to discuss the effects that these attitudes held by "the West" have influenced the uptake of GM crops by Africa and "the Rest." I will then investigate some of the myths that have been used to turn politicians, decision-makers, and inhabitants of such countries against GM crops, and to consider the importance of communication. As it is necessary to understand why and how certain countries "got it right" and are currently growing GM crops successfully, the last section deals with these issues. The conclusion points to the necessity for countries to learn from mistakes made in the past as we enter the era of new technologies such as genome editing.

Keywords: developed, developing, myths, regulations

1. Introduction

This chapter is based on a book I recently wrote, entitled: GM Crops and the Global Divide [1]. In the preface to that book, I talked about the need for the bridge across the agricultural genetic divide between African countries and those in the developed world to be crossed. This divide separates the use of genetically improved varieties available in the developed world from those being used by resource-poor farmers in Africa. Here, I consider how attitudes to GM crops found in countries in the "West," especially in the European Union (EU), have had a negative effect on their uptake in Africa and other developing countries. As much of the West's attitude is based on myths and disinformation (when untruths are deliberately spread as opposed to misinformation, which could be based on ignorance of the truth), I have included a section on how important truthful communication is in this debate. I then go on to discuss the countries that "got it right" and how this came about. I sincerely hope that we can learn from such success stories and use them to guide regulations going forward into new technologies such as gene editing, which can be enormously helpful in bringing about improved food security in countries that sorely need this.

2. The West's stand on GM crops

The global area where GM crops have been planted has grown from an initial 1.7 million hectares from the time they were first commercialized in 1996 to over 190 million in 2019. However, the current top 10 growers are the USA, Brazil, Argentina,

Canada, India, Paraguay, China, Pakistan, South Africa, and Bolivia. The only EU country to appear in the list of the top 21 is Spain, coming in at number 17 [2]. This is clearly a reflection of each country's approach to regulations of GM crops. For instance, Canada's regulations are based on a scientific analysis of the traits and whether they are beneficial. No attention is paid to the methods by which such traits had been achieved [3]. Many of the top growing countries took much the same approach as that taken by Canada. By contrast, in the EU, the traits themselves are of little consequence and the methods used in developing the GM crop are of paramount importance [3]. How did these differences in attitude come about?

The development of GM crops in Europe occurred at much the same time as initial steps were being taken to integrate national food safety systems into the European Food Safety Authority (EFSA). This was politically sensitive because individual countries in the EU were losing some of their influence over home-based regulations. In my book GM Crops and the Global Divide [1], I postulate that the US biotechnology industry blustered its way into the EU, hoping to sell their GM crops to European farmers in this already somewhat hostile regulatory environment. As pointed out by Wesseler and Kalaitzandonakes [4] "Never before has a new technology in the field of agriculture been so emotionally debated among stakeholders." I might add, however, that, according to the latest analysis carried out by the International Service for the Acquisition of Agribiotech Applications [2], GM crops have increased about 122-fold from 1996 to 2019 making biotechnology the fasted adopted crop technology in the world.

Many countries, including the EU, cite the precautionary principle as a reason for not allowing the cultivation of GM crops. This principle, in essence, states that if an action has a suspected risk of causing harm to the public or to the environment, the burden of proof (that it is or is not harmful) falls on those taking the action. A major problem with this is the difficulty of proving it negative, thus establishing evidence of the absence of danger is difficult. Indeed, on the basis of this principle why are cars allowed on the roads?

It should also be borne in mind that there is a great difference between the blanket statement of "risks of GM crops" and the specific statement of "risks of approved GM link MON810," or "risks of insect resistant soybeans in Argentina." In addition, should not there be an overriding proviso when benefit-risk ratios are taken into account, such as is obviously the case for cars? Here, again, the West might well say (as they often do): "We have enough food therefore we don't need food derived from GM crops." On the contrary, the Rest might answer (which I wish they would do more vociferously): "We need any technology that puts more food on our tables." Perhaps decision-makers in the EU could benefit by spending time living in rural India, Paraguay, or Bolivia (numbers 5, 6, and 10 of ISAAA's list [2]) before making up their minds on the usefulness or not of GM crops.

3. The West versus Africa

The only countries in Africa that are currently growing GM crops commercially are South Africa, Sudan, and very recently, Eswatini and Nigeria. On June 22, 2021, Kenya announced that it had approved the environmental release of GM cassava resistant to cassava brown streak disease, which had been developed by the Kenya Agricultural and Livestock Research Organization (KALRO). This paves the way for national performance trials before it can become commercialized.

Why is this and why are not more African countries growing GM Crops? It is not as if South African farmers have had bad experiences, especially when growing GM white maize which can be eaten by some citizens up to three times per day. In a recent article entitled Economic and Ecosystem Impacts of GM Maize in South Africa [5], the authors state that the key benefits of growing GM white maize were estimated to amount to US$5 million from 2001 to 2018, with lower pesticide requirements compared to convention white maize. In 2017, South Africa produced approximately 1.1 million hectares of GM maize varieties for direct human consumption, representing an 85% adoption rate. In light of food insecurity in African countries, which will only become worse with climate change, why do not grow more GM crops?

To understand the influence that Europe could be having in Africa, it is important to understand the role that Europe plays in both the economy and mindset of many African countries and their leaders. A statement by the European Commission reads: "in an ever-changing world, one thing is for sure: Africa and Europe will remain each other's closest neighbours. Africa's 54 countries and the European Union's 28 Member States have a shared neighbourhood, history and future" [6].

Moreover, Africa's farm exports to Europe are six times as large as exports to the United States, so it is European consumers' taste and European regulatory systems that Africans most often must adjust to. In addition, Europe provides three times the funding for the United Nations Environmental Program (UNEP) which, together with the Global Environment Facility, provided assistance to African regulatory authorities. Therefore, Europe could influence organizations to adopt EU-style restrictions on GM crops, and the EU has been waging a war on GMO foods for decades [7].

What is Africa losing by not planting GM crops? Justus Wesseler, an agricultural economist from Wageningen University in the Netherlands, published an article in 2017 in which he and his colleagues considered the cost of not growing three GM crops. These were disease-resistant cooking bananas (plantains), and insect-resistant maize and cowpea [8]. They estimated that in the past decade, between 440 and 4000 lives could have been saved in Kenya, while in Uganda the potential estimate was between 500 and 5500.

Of course, Europe is not the only entity trying to stop Africa from growing GM crops. Western non-governmental organizations (NGOs), such as Greenpeace, Friends of the Earth, GeneWatch UK, ActionAid, and GM Freeze, are all cited in an article written in 2017 by Margaret Karembu, Director of the ISAAA AfriCenter, soon to become BioTrust AfriCenter [9]. The title of her article is "How European-based NGOs block crop biotechnology adoption in Africa." She grew up in rural Kenya when her family struggled to put food on the table. She now realizes that the subsistence farming practiced by her family is what the "greens" in Europe and elsewhere in the West call "agro-ecology family farming." Although farming practices in Africa are beginning to modernize, this is being undermined by such organizations. She gives the example of the adoption by the European Parliament in June 2016 of a report by the "New Alliance for Food Security and Nutrition," which stated that any support to African agriculture should be confined to such "agro-ecology family farming level." The report was passed by 577 to 24.

Africa is not the only continent that has been the object of the anti-GMO lobby. The next section will look at how other continents and countries have been affected.

4. The West versus the Rest

What effects have the West had on other "developing" countries? I used quotation marks as one hopes that all countries are developing, although in the minds of most people, "developing" countries are those not as economically advanced as the

"West". Therefore, in this section, I will look at the effects that the West has had on eggplants (brinjals, aubergines, or talong) in the Philippines and Bangladesh, and on Golden Rice in Asia in general.

Eggplants (*Solanum melongena* L.) are among the most important, inexpensive, and popular vegetables grown and consumed in Asia. In the Philippines, for instance, they account for more than 30% of the total volume of vegetables produced in the country [10]. The problem, however, is that they are susceptible to infestation by the eggplant fruit-and-shoot borer (EFSB; *Leucinodes orbonalis* Guenée), and farmers use chemical insecticides to control these pests. Indeed, farmers in the Philippines can apply these chemicals 20–72 times during the 5- to 6-month-long cultivation season, often resulting in skin irritation, redness of eyes, muscle pains, and headaches in farmworkers [11]. As there are no conventionally bred-resistant varieties, the Maharashtra Hybrid Seeds (Mahyco) developed GM resistance that gave 98–99% damage loss [10]. At first, the Supreme Court of the Philippines placed a permanent injunction on field trials, but this was later overturned.

The situation in Bangladesh is very different. Mahyco, working with the Bangladesh Agricultural Research Institute, developed nine varieties of Bt eggplant that eliminated the need to spray for EFSB. Four of these varieties received regulatory approval in 2013 and were grown by 20 farmers in 2014. Today, more than 27,000 farmers in Bangladesh grow Bt eggplant, and there are indications that more farmers are eager to reap the benefits of these improved varieties.

https://www.agrilinks.org/post/bt-eggplant-adds-revenue-safety-farmers-bangladesh.

In a recent article [12], the authors found that Bt eggplant varieties had a 19.6% higher average yield and 21.7% higher revenue. This amounted to $664 more income per hectare, a princely sum for resource-poor farmers in Bangladesh.

The study also found that Bt eggplant sold at the local markets, either to wholesalers or direct to consumers, fetched a higher price than non-Bt eggplant. Some buyers were prepared to pay higher prices for Bt eggplant because the fruit was less damaged than non-Bt eggplant.

What were the reasons for this success in Bangladesh? One is probably the partnership between Mahyco, the United States Agency for International Development (USAID), the Indian-based Sathguru Management Consultants who helped with technology transfer and innovation advice, and Cornell University. The group is now called the South Asia Eggplant Improvement Partnership (SAEIP) and they designated the Bangladesh Agricultural Research Institute (BARI) as the lead organization in producing and distributing Bt eggplants to farmers [12].

Another important factor was that the four Bt eggplant lines released were not hybrids, so farmers could save seed. In addition, BARI provided farmer training, explaining the importance of planting refuge non-Bt eggplants around the Bt eggplant plots to prevent the build-up of weeds resistant to Bt. The satisfaction of the farmers with their crops prevented the anti-GMO lobby, which was very active in the early days of the rollout of the crop, from turning away government support.

As mentioned above, the other case I am going to look at is that of Golden Rice. In many parts of Asia, rice is eaten almost every day, in some countries accounting for 70–80% of an individual's calorie intake [13]. Unfortunately, as rice is prepared by the removal of the husk and aleurone layer to prevent the grains from becoming rancid during storage, micronutrients, including vitamin A, are removed. This can lead to vitamin A deficiency (VAD), which has been estimated to kill approximately 670,000 children under the age of 5 years every year [14]. In addition, VAD can cause an additional 500,000 cases of irreversible blindness [15].

To address this problem, in the early 1990s two scientists from the Swiss Federal Institute of Technology (ETH) in Zurich, Ingo Potrykus and Peter Beyer, began to develop a variety of rice that would contain vitamin A. They introduced two genes, one from maize and the other from a very commonly ingested soil bacterium, *Erwinia uredovora*, which together produced lycopene. The rice converts this to β-carotene, which gives the rice a golden color and is converted to vitamin A when ingested. Hence the name Golden Rice. https://archive.gramene.org/newsletters/rice_genetics/rgn18/c41.html

This variety has been the subject of many attacks by anti-GMO activists, spearheaded by Greenpeace. Incensed by this opposition that has been ongoing for many years, some 150 Noble Laureates wrote an open letter to the leaders of Greenpeace, as well as to the United Nations and governments around the world urging "Greenpeace and its supporters to re-examine the experience of farmers and consumers worldwide with crops and foods improved through biotechnology.... and abandon their campaign against GMOs in general and Golden Rice in particular" [16].

Although Golden Rice has been approved for use in Australia, New Zealand, Canada, and the United States, these are the countries that hardly need it. At last, in December 2019, the Philippines approved its use in food, feed, and for processing [17], and finally, on July 28, 2021, Golden Rice was approved for commercial planting by the Philippine Department of Agriculture.

https://www.thepigsite.com/news/2021/07/philippine-department-of-ag-gives-nod-to-nutritious-gmo-golden-rice.

5. How to bust myths and the importance of communication

A number of myths have grown over the years regarding GMOs and GM crops. It is, of course, important to use scientific facts to correct these myths, but it is equally important to understand that many people who believe them may be doing so to reinforce some beliefs that they hold. Simply giving them the facts may not be enough to dispel such myths, hence it is important to understand what lies behind their adherence to them. For instance, if a person is against the role of multinational companies, which they believe are monopolizing the production of GM crops, it is essential to present the facts as they relate to this issue. Above all, it is necessary to gain the trust of such opponents of GM crops before you have any hope of convincing them otherwise.

People often tend to base their decisions on opinions and values and then look for facts that support these. To counter this, you will have to obtain their trust, and one way to do this is to state that there are aspects of GM crops that you find problematic. For instance, you can acknowledge that the overuse of a single herbicide such as Roundup can lead to the development of herbicide-resistant crops. However, you should point out that this is not the fault of the technology but the use of such technology. So, let us look at a number of these myths.

5.1 Superweeds

This term is an emotive one aimed at inspiring fear. However, "superweeds" are no different from the herbicide-resistant weeds found in fields of conventional crops, which farmers have been dealing with for many years. The term also implies that there is no herbicide that can kill such weeds, which is patently untrue. In contrast, the development of "super bugs," referring to human bacterial infections that are resistant to many or even to all known antibiotics, are real threats. There are, indeed, multidrug-resistant (MDR) and extremely drug-resistant (WDR) tuberculosis strains that are presenting formidable challenges to treatment [18].

That said, Roundup-resistant weeds are, indeed, a growing problem that needs to be addressed. However, to put it into perspective, by 2014, weeds have become resistant to 152 different herbicides, emphasizing the importance of managing weeds in a more integrated and sustainable manner [19]. Indeed, in the case of Roundup-resistant weeds, farmers are their own worst enemies—the more they continue to use this herbicide without rotation, the greater chance there will be for the development of resistant weeds.

Another common myth is that poor farmers in Africa have to buy maize seeds every year and cannot save seeds. However, since the advent of hybrid seeds in the 1930s, farmers who plant them have to buy seeds every year. This is because of the way in which hybrids are bred. Specific male and female lines that have been bred so that their offspring (hybrids) have advantageous traits such as high yield. If farmers plant their own seeds, the offspring will be a scramble of traits as their parental genes are randomly inherited, losing the "hybrid vigor."

A complaint, not a myth, that is often used against GM crops, is that they are in the hands of the multinationals who, and this is a myth, are plotting to control the food supply of developing nations. Why do multinationals produce most of the world's GM crops? This is, in fact, due to the anti-GMO lobby. They have stirred up such fears of harm to humans, animals, and the environment that regulations imposed by governments have become enormously expensive. As a result, only multinationals with deep pockets can afford to comply with these regulations.

I have seen this problem at first hand. Some years ago, colleagues and I developed GM maize resistant to the African endemic maize streak virus [20]. Our private sector partner was the South African seed company, Pannar Seed, who simply could not afford the costs involved in carrying out field trials. Some years later, they were taken over by the multinational seed company, Pioneer, but they too were unable to undertake such trials. Their reason was that MSV-resistant maize would only benefit African farmers who were too poor to recoup the costs of field trials. The seeds remain in the freezers at the University of Cape Town and Pannar Seed.

6. Countries that got it right and why

What do I mean by a country that got it right? In a nutshell, such a country needs to have a government that is supportive of innovations and new technologies that can improve agricultural production and make the lives of farmers more profitable and less stressful. It should have a regulatory system in place that is flexible, operates on a case-by-case system, and whose decisions are based on science. It is extremely important that these regulations should not erect barriers to the development and implementation of GM crops. The government should encourage private enterprises to develop and commercialize such crops, and should also support public enterprises such as universities, technical colleges, research institutes to conduct research that could be commercialized in public/private partnerships. I will now give some examples of countries that got it right and how they did this.

6.1 South Africa

Farmers in South Africa started to plant GM crops commercially as early as 1998. In 2019, it was number 8 on the list of countries planting the highest number of GM crops with 2.7 million hectares of maize, soybeans, and cotton. Maize, at 72%, accounts for the majority of these, with soybeans at 27% and cotton a mere

1%. Approximately 85% of white maize, used for human consumption, was GM. Indeed, white maize is often consumed three times a day by many inhabitants [2].

In a recent study of the economic and ecosystem impacts of GM maize in this country, the authors found that white maize food security was improved as, on average, 4.6 million additional rations were added annually. In addition, the environmental impacts per hectare of GM versus non-GM maize were decreased by US$0.34 per hectare, or US$291,721 annually. Decreases in pesticides accounted for the majority of the estimated US$5 million benefits from 2001 to 2018. The authors speculate that "as we face a hotter and drier future, agricultural technologies such as GM may be one of the most salient ways to combat food insecurity while simultaneously reducing the environmental impact of agricultural production" [5].

6.2 Canada

Canada regulates products derived from biotechnology processes as part of its existing regulatory framework for "novel products." The focus is on the traits expressed in the products and not on the method used to introduce those traits...Advertising or labeling the presence of GMOs in particular food is voluntary unless there is a health or safety concern.

Thus, reads the introduction to the Law Library of Congress's [21] article entitled "Restrictions on Genetically Modified Organism Canada." In keeping with this approach to GMOs, Canada, the world leader in canola production, was the first country to commercialize herbicide-tolerant (HT) varieties of this crop in 1996. By 2019, 82% of its country's soybeans were HT, 90% of its maize crop was either both HT and insect-resistant (IR) or either one, and 95% of its canola was HT [2].

While most countries growing GM crops concentrate on maize, soybeans, cotton, or canola, which are either HT or IR or both. However, Canada has recently commercialized three new crops. The first was HT low-lignin alfalfa, which makes it more digestible to livestock [22]. This also allows farmers to delay harvest by up to 10 days in order to obtain greater yields without losing quality [2]. This was developed by members of a partnership between Forage Genetics International, the Noble Foundation, the US Forage Research Centre, together with scientists from the universities of Wisconsin, Minnesota, and the University of California, Dave—a great example of public/private collaboration [23].

The second was the Innate® potato developed by the JR Simplot company. It has decreased levels of reducing sugars, reduced acrylamide potential (by reducing asparagine), and black spot bruising tolerance [24]. This crop has been further improved by protection against the late blight pathogen, which could result in up to a 50% reduction in fungicide application annually [2]. Reduced asparagine leads to lower levels of acrylamide (potentially carcinogenic) that can accumulate when potatoes are cooked at high temperatures, for instance over open flames.

The third new crop is the Arctic Apple produced by the Canadian firm Okanagan Specialty Fruits Inc. The development of these crops is a good example of how a country "got it right" as private companies, or public/private partnerships, are clearly encouraged by the fact that if their products are an improvement on what is currently available, Canada will allow their commercialization as long as there are no health or safety issues involved [2].

6.3 Argentina

GM crops are regulated in Argentina under the general Law on Seeds and Phytogenetic Creations. This law aims to promote the development and production of modern biotechnology as it grants tax incentives to research and production projects that meet safety and health standards. This forward-looking approach assures farmers that the seed they acquire meets identity and quality standards, while the intellectual property of innovators is protected [25].

Argentina was among the first countries to plant GM crops commercially, with HT soybeans being introduced in 1996; in 2019, it was the third largest grower of such crops [2]. One of the major reasons for this uptake is that farmers can plant two crops per year, partly due to the reduced tilling required with HT crops as this reduces the production time. This is because the farmers grow conventional varieties of these crops, they need to till the soil before planting in order to allow weeds to grow. These are then killed with herbicides, many of which are not biodegradable. Farmers must, therefore, wait until the herbicide has dissipated before the crop can be planted; otherwise, the residual herbicide remaining in the soil will kill the crop. During this time, much of the topsoil may be lost due to wind erosion. Argentinian farmers learned early on that they could spray the fields of HT soybeans with minimal tilling and when it was convenient for them as only the weeds would be killed. They also noted that less topsoil was being lost. https://www.isaaa.org/resources/publications/pocketk/57/default.asp.

One of the other things that Argentina "got right" was that its government recognized that GM crops have the potential to increase agricultural output. In March 2017, the agriculture minister, Dr. Luis Miguel Etchevehere, said that the promotion of GM crops was designed to increase the "leadership of our country in the development of agricultural biotechnology" [26]. They also implemented regulatory measures aimed at speeding up the approval of new GM crops, resulting in the time taken for such assessments dropping from 42 to 24 months [26]. This has resulted in the country approving HarvXtra® Alfalfa, which contains less lignin thus improving its digestibility for livestock. In addition, Argentina is the first country to approve drought-tolerant GM wheat [27]. Both crops were developed by BiOceres, a local agri-industrial company.

6.4 Brazil

In September 2003, Brazil decided to allow farmers to grow GM soybeans for a 1-year period. As Bob Callanan, head of the pro-GM American Soybean Association, said: "We have long been frustrated by Brazil growing illegal GM seeds" [28]. These seeds were being brought over the border by farmers seeing how much better off farmers were in Argentina due to their growing GM varieties. Far from limiting its planting of GM crops to a single year, Brazil's farmers have grown them continuously and, in 2019, they were growing the second largest area of these crops, 52.8 million hectares, with the USA, at 71.5, growing the largest [2].

6.5 China

Since 1997, China has approved 64 GM crop events, including canola, cotton, maize, papaya, petunia, poplar, rice, soybeans, sugar beets, sweet pepper, and tomatoes [2]. This sounds impressive, so why then in 2019 is China planting only 3.2 million hectares, the seventh in the list of countries planting the highest area to GM crops? [2] The answer may lie in the country's regulators being risk-averse. As in many countries in the 1900s, Greenpeace was a major player in China and the specter of risk was high on its agenda. Thus, after its initial early entry into GM crops, China, hopefully for the moment, has dropped by the wayside [29].

6.6 Burkina Faso

In 2008, farmers in Burkina Faso, one of Africa's largest cotton producers, began to plant IR cotton commercially. By 2014, about 74% of cotton grown in the country was IR, grown by some 140,000 smallholder farmers. Advantages included a 20% yield increase, a reduction in insecticide use of about 67%, and an estimated profit increase of US$64 per hectare, despite the increase in the cost of seed [30].

However, although the farmers were happy, the cotton ginning companies were not. The cotton had shorter staples and lower lint quality undermined their profit. As a result, the cotton companies, which also control the provision of seed to the farmers, unilaterally phased out GM cotton [31]. This is an example of the importance of having all involved in the crop production and processing involved in the decisions taken as to which varieties of cotton should be converted into GM, in this case by inserting the *Bt* gene.

7. Conclusions

Why is it that people in the West, particularly those in the European Union but also in the Nordic countries and parts of the USA, are so against GM crops? Since 1996 when such crops became available commercially, not a single proven case of ill health related to their consumption by either humans or animals is registered. Every major regulatory body in the world has concluded that GM crops are as safe for consumption as conventional crops, whether organic or not [3].

Is it possible that there are vested interests involved in this antagonism? One of the most prominent organizations lobbying against GM crops is Greenpeace, which receives funding from other anti-GMO bodies such as the Tides Foundation. https://www.tides.org/project/grantee/greenpeace-canada/.

Another organization working against GM crops is the Norwegian Institute of Gene Ecology (GenØk), which has been fiercely opposed to this technology since it was founded in 1998. Their staffs travel widely promoting perceived risks associated with this technology. They also hold conferences such as the one in 2003 with the inflammatory title: "Regulating a privatized genetic industry which has the potential to destroy the future." [32].

http://fafdl.org/blog/2016/10/14/how-norway-became-an-anti-gmo-powerhouse/.

Looking to the future, will the public and regulators around the world accept the newer technology of gene editing? [33] The potential that this technology has for improving crops and food sustainability is enormous. Let us learn from the mistakes made over GM crops and not repeat them. In particular, let us be aware that decisions made in the West can have a huge impact on the actions taken by the Rest.

Author details

Jennifer Thomson
University of Cape Town, South Africa

*Address all correspondence to: jennifer.thomson@uct.ac.za

References

[1] Thomson JA. GM Crops and the Global Divide. Australia: CSIRO Publishing; 2011

[2] ISAAA. Global Status of Commercialized Biotech/GM Crops in 2019: Biotech Crops Drive Socio-Economic Development and Sustainable Environment in the New Frontier. ISAAA Brief No. 55. Ithaca, NY: ISAAA; 2019

[3] Smyth SJ, McHughen A. Regulation of genetically modified crops in USA and Canada: Canadian overview. In: Wozniaö C, McHughen A, editors. Regulation of Agricultural Biotechnology: The United States and Canada. New York, NY: Springer Publishers; 2012. pp. 15-34

[4] Wesseler J, Kalaitzandonakes N. Present and Future EU GMO policy. In: Oskam A, Meesters G, Silvis H, editors. EU Policy for Agriculture, Food and Rural Areas. 2nd ed. Wageningen, Netherlands: Wageningen Academic Publishersroch; 2011. pp. 403-413

[5] Ala-Kokko K, Nalley LL, Shew AM, Tack JB, Chaminuka P, Matlock MD, et al. Economic and ecosystem impacts of GM maize in South Africa. Global Food Security. 2021;**29**:100544. DOI: 10.1016/j.gfs.2021.100544

[6] European Commission. Africa–EU Continental Cooperation. 2019. Available from: https://ec.europa.eu/europeaid/regions/africa/africa-eu-continental-cooperation_en

[7] Paarlberg RL. Starved for Science: Keeping Biotechnology Out of Africa. Cambridge, MA: Harvard University Press; 2008

[8] Wesseler J, Smart RD, Thomson JA, Zilberman D. Foregone benefits of important food crop improvements in sub-Saharan Africa. PLoS One. 2017;**27**:e0181353. DOI: 10.1371/journal.pone.0181353

[9] Karembu M. How European-based NGOs block crop biotechnology adoption in Africa. Reposted from Genetic Literacy Project. 2017. Available from: http://africenter.isaaa.org/european-based-ngos-block-crop-biotechnology-adoption-africa

[10] Hautea DM, Taylo LD, Masanga AP, Sison ML, Narciso JO, Quilloy RB, et al. Field performance of Bt eggplants (*Solanum melongena* L.) in the Philippines: Cry1Ac expression and control of the eggplant fruit and shoot borer (Leucinodes orbonalis Guenée). PLoS One. 2016;**11**(6):e0157498

[11] Del Prado-Lu JL. Insecticide residues in soil, water, and eggplant fruits and farmers' health effects due to exposure to pesticides. Environmental Health and Preventive Medicine. 2015;**20**(1):53-62

[12] Shelton AM, Hossain MJ, Paranjape V, Azad AK, Rahman ML, Khan AS, et al. Bt eggplant project in Bangladesh: History, present status, and future direction. Frontiers in Bioengineering and Biotechnology. 2018;**6**:106

[13] Bouis H, Islam Y. Biofortification: Leveraging agriculture to reduce hidden hunger. In: Fan S, Pandya-Lorch R, editors. Reshaping Agriculture for Nutrition and Health. Washington, DC: International Food Policy Research Institute; 2012. pp. 83-92. Available from: https://books.google.co.za/books?hl=en&lr=&id=gVQZXr38z5sC&oi=fnd&pg=PA83&dq=Bouis+H+and+Islam+Y+(2012)+Biofortification:+Leveraging+agriculture+to+reduce+hidden+hunger&ots=0Dg5ZY4nr8&sig=Tz4z6Z0ILtGxjBgtWPIdRDjXpQk#v=onepage&q&f=false></edb

[14] Black RE, Allen LH, Bhutta ZA, Caulfield LE, De Onis M, Ezzati M, et al. Maternal and child undernutrition: Global and regional exposures and health consequences. The Lancet. 2008;**371**(9608):243-260

[15] Humphrey JH, West KP Jr, Sommer A. Vitamin A deficiency and attributable mortality among under-5-year-olds. Bulletin of the World Health Organization. 1992;**70**(2):225

[16] Roberts RJ. The Nobel Laureates' campaign supporting GMOs. ScienceDirect. 2018;**3**:61-65. DOI: 10.1016/j.jik.2017.12.006

[17] Dubock A, Potrykus I, Beyer P. Philippines is first! Long-delayed Vitamin A-enhanced Golden Rice greenlighted, bucking activist opposition. In: Genetic Literacy Project, 18 December 2019. 2019. Available from: https://geneticliteracyproject.org/2019/12/18/philippines-is-first-long-delayed-vitamin-a-enhanced-golden-rice-greenlighted-bucking-activist-opposition

[18] Seung KJ, Keshavjee S, Rich ML. Multidrug-resistant tuberculosis and extensively drug-resistant tuberculosis. Cold Spring Harbour Perspectives in Medicine. 2015;**5**(9):a017863. DOI: 10.1101/cshperspect.a017863

[19] Heap I. Herbicide resistant weeds. In: Pimentel D, Peshin R, editors. Integrated Pest Management. Dordrecht: Springer; 2014. pp. 281-301

[20] Shepherd DN, Mangwende T, Martin DP, Bezuidenhout M, Kloppers FJ, Carolissen CH, et al. Maize streak virus-resistant transgenic maize: A first for Africa. Plant Biotechnology Journal. 2007;**5**(6):759-767

[21] Law Library of Congress. Restrictions on Genetically Modified Organisms. Canada: Law Library of Congress; 2012. Available from: https://www.loc.gov/law/help/restrictions-on-gmos/canada.php

[22] Guo D, Chen F, Wheeler J, Winder J, Selman S, Peterson M, et al. Improvement of in-rumen digestibility of alfalfa forage by genetic manipulation of lignin O-methyltransferases. Transgenic Research. 2001;**10**(5):457-464

[23] Forage Genetics International. Get Xtra Control and Xtra Quality where It Counts. 2019. Available from: https://www.foragegenetics.com/Products-Technologies/HarvXtra-Alfalfa

[24] Halterman D, Guenthner J, Collinge S, Butler N, Douches D. Biotech potatoes in the 21st century: 20 years since the first biotech potato. American Journal of Potato Research. 2016;**93**(1):1-20

[25] Rodriguez-Ferrand G. Spain: Parot Doctrine after the ECHR decision. In: The Law Library of Congress. 2014. Available from: https://www.loc.gov/law/help/parot-doctrine/spain-parot-doctrine.pdf

[26] Cerier S. Argentina and GMOs: Exploring the nation's long relationship with biotech crops. In: Genetic Literacy Project, 6 September 2018. 2018. Available from: https://geneticliteracyproject.org/2018/09/06/argentina-and-gmos-exploring-the-nations-long-relationship-with-biotech-crops

[27] Sheridan C. Argentina first to market with drought-resistant GM wheat. Nature Biotechnology. 2021;**39**:652. DOI: 10.1038/s41587-021-00963-y

[28] Vidal J, Chetwynd G. Brazil agrees to grow GM seed. The Guardian. 2003, 2003. Available from: https://www.theguardian.com/science/2003/sep/26/gm.food

[29] Cao C. GMO China: How Global Debates Transformed China's

Agricultural Biotechnology Policies. Contemporary Asia in the World. New York, NY: Columbia University Press; 2018

[30] James C. Global status of commercialized biotech/GM crops: 2014'. ISAAA Brief No. 49. Ithaca, NY: ISAAA; 2014

[31] The Conversation. Lessons to be learnt from Burkina Faso's decision to drop GM cotton. The Conversation Australia, 2016. 2016. Available from: https://theconversation.com/lessons-to-be-learnt-from-burkina-fasos-decision-to-drop-gm-cotton-53906

[32] Feng Z, Zhang B, Ding W, Liu X, Yang DL, Wei P, et al. Efficient genome editing in plants using a CRISPR/Cas system. Cell Research. 2013;**23**(10): 1229-1232

[33] Langberg L, Heggdal Ø. GenØk: How Norway came to revile GMOs. In: Genetic Literacy Project, 18 October 2016. 2016. Available from: https://geneticliteracyproject.org/2016/10/18/genok-norway-came-revile-gmos/> </eref